青杨雌雄植株及其交互嫁接
对干旱和铅胁迫的生理生态响应及机制

韩 颖 编著

项目策划：蒋 玙 龚娇梅
责任编辑：蒋 玙
责任校对：龚娇梅
封面设计：墨创文化
责任印制：王 炜

图书在版编目（CIP）数据

青杨雌雄植株及其交互嫁接对干旱和铅胁迫的生理生态响应及机制 / 韩颖编著． — 成都：四川大学出版社，2021.9
ISBN 978-7-5690-5062-2

Ⅰ．①青… Ⅱ．①韩… Ⅲ．①青杨－雌雄同株－生态特性－研究 Ⅳ．①S792.113.02

中国版本图书馆CIP数据核字（2021）第206481号

书　名	青杨雌雄植株及其交互嫁接对干旱和铅胁迫的生理生态响应及机制
编　著	韩　颖
出　版	四川大学出版社
地　址	成都市一环路南一段24号（610065）
发　行	四川大学出版社
书　号	ISBN 978-7-5690-5062-2
印前制作	四川胜翔数码印务设计有限公司
印　刷	四川五洲彩印有限责任公司
成品尺寸	148mm×210mm
印　张	4
字　数	103千字
版　次	2021年10月第1版
印　次	2021年10月第1次印刷
定　价	42.00元

版权所有 ◆ 侵权必究

◆ 读者邮购本书，请与本社发行科联系。
　电话：（028）85408408/（028）85401670/（028）86408023　邮政编码：610065
◆ 本社图书如有印装质量问题，请寄回出版社调换。
◆ 网址：http://press.scu.edu.cn

四川大学出版社
微信公众号

前　言

在中国，干旱和铅污染逐渐成为限制森林生态系统生产力的重要因素，尤其对雌雄异株森林生态系统构成了严重的威胁。由于雌性植株主要负责产生种子和果实，雄性植株主要负责传播花粉，因此，二者担负的繁殖成本存在显著差异。雌雄异株木本植物的雌性通常比雄性分配更多的资源用于繁殖，而较少的资源用于维持生长和抵抗逆境，进而雌雄个体对逆境胁迫产生了性二态性响应。本书以青杨雌雄植株为模式植物，采用交互嫁接技术，探讨了在干旱和重金属铅的双重胁迫下雌雄幼苗的形态生长特征、生理生化响应、细胞超微结构变异以及对铅污染土壤的修复潜力，揭示了青杨雌雄植株对干旱胁迫的差异响应机理。本书可为雌雄异株植物应对环境变化响应研究积累重要基础资料，也可为重金属污染样地修复树种的选择提供重要借鉴。

本书共包含 3 篇 7 章。第 1 篇为序篇，主要介绍一些常见术语，如雌雄异株、性二态性、交互嫁接、光合速率。第 2 篇为试验方法篇，主要介绍本书研究常用的方法与技术原理，包括杨树的嫁接方法、气体交换测量方法、生长参数及生理指标测量方法、试验设计与统计分析方法。第 3 篇为研究结果与分析篇，包括青杨雌雄植株对干旱胁迫、铅胁迫及交互胁迫的响应差异、青杨雌雄交互嫁接植株对干旱胁迫的响应差异。

虽然学者对雌雄异株植物的逆境响应已经开展了大量研究，

但这方面的研究远没有结束，还有很多机制需深入挖掘。此外，由于作者水平有限，书中如有错误，恳请读者批评指正。

特别感谢我的博士生导师李春阳研究员，是他带领我进入这样有趣的学术领域，并把我培养成具有独立开展研究能力的学者；同时也感谢所有关心和帮助我的人。

韩　颖

2021 年 5 月 5 日

目　　录

序　篇

第1章　干旱和铅胁迫下雌雄异株植物的性二态性……（ 3 ）
　§1.1　常用术语……………………………………（ 5 ）
　§1.2　雌雄异株植物的性二态性研究进展 …………（ 6 ）
　§1.3　植物对干旱和铅胁迫的生理生态响应及研究进展
　　　　………………………………………………（ 7 ）
　§1.4　嫁接植物对非生物胁迫的响应 ………………（ 15 ）

试验方法篇

第2章　材料与试验处理方法………………………（ 27 ）
　§2.1　植物材料的采集和培养 ………………………（ 27 ）
　§2.2　试验处理方法 …………………………………（ 28 ）

第3章　气体交换测量方法…………………………（ 30 ）
　§3.1　光合作用分析 …………………………………（ 30 ）
　§3.2　叶绿素荧光分析 ………………………………（ 32 ）

第4章 生长参数及生理指标测量方法 (35)
§4.1 生长和形态参数测量 (35)
§4.2 生理指标测定 (36)

第5章 试验数据统计分析 (38)
§5.1 干旱和铅试验数据统计分析 (38)
§5.2 干旱和嫁接试验数据统计分析 (38)

研究结果与分析篇

第6章 青杨雌雄植株对干旱胁迫、铅胁迫及交互胁迫的响应差异 (43)
§6.1 主要研究结果 (43)
§6.2 讨论 (64)
§6.3 小结 (70)

第7章 青杨雌雄交互嫁接植株对干旱胁迫的响应差异 (72)
§7.1 主要研究结果 (72)
§7.2 讨论 (81)
§7.3 小结 (85)

主要结论 (86)

参考文献 (88)

序篇

第1章　干旱和铅胁迫下雌雄异株植物的性二态性

随着全球气候变暖和温室效应加剧，以及全球生态环境不断恶化，干旱已经成为威胁人类生存和发展的主要灾害。目前全球森林系统正面临着极大的"干旱致死"风险。铅作为对环境威胁最大的污染物之一，在土壤中长期存在，并且对植物、动物有很强的毒性。中国拥有丰富的铅矿资源，是世界上最大的铅生产基地，中国自20世纪以来增加的铅矿开采已经明显增加了空气铅排放和土壤铅污染。铅既不参与植物体的结构组成，又不参与细胞的代谢活动，但易被植物根部吸收，从而对植物产生毒性，是目前中国乃至世界土壤－植物生态系统中主要的重金属污染物，对森林生态系统构成一定的威胁。空气铅和土壤铅可以改变森林生态系统结构、动力学以及多样性，影响铅敏感树种的生长和发育，并减少其种群数量，甚至导致灭绝和消失。研究表明，在全球生态环境变化的大背景下，未来干旱和重金属这两个胁迫因子的结合会更加广泛。在中国，干旱分布区和铅矿分布区大部分交叠在一起，这两种环境胁迫因子频繁共存于森林生态系统，对森林生态系统造成了严重的威胁。尽管目前许多被子植物对单独的干旱胁迫以及单独的铅胁迫的生理、生态和生化响应已逐步得以报道，但是有关重金属和干旱交互作用这一主题的知识相当匮乏，其中关于干旱和铅的交互研究也很少见。

值得注意的是，在植物界30多万种被子植物中，还存在4%~10%的雌雄异株植物，据Renner & Ricklefs（1995）对24万种被子植物的调查表明，雌雄异株植物占959个属，共计14620种。而在森林生态系统中，有16个科为雌雄异株或部分雌雄异株。常见树木如千年桐、猕猴桃、银杏、香榧、桑树、白蜡树、花椒，棕榈、杨树和柳树都属于雌雄异株植物。雌雄异株植物由于存在不同的生殖资源投入问题，在长期的进化过程中，一些物种的雌雄株性比偏离1∶1。在资源匮乏的环境中，雌雄异株植物种群性比偏雄性，相反性比偏雌性。全球变化所导致的资源可利用性和分布格局的改变以及环境污染的加剧可能会进一步增大雌雄异株植物的性比偏离程度，并导致雌雄异株植物繁殖成功率急剧下降。其中水资源的匮乏和铅污染可能会导致持续生境片段化，最终导致物种多样性丧失。而相比其他物种，雌雄异株植物种群大小和结构更易改变，对生境片段化更加敏感。因此，开展雌雄异株树种对干旱和重金属铅交互作用的研究，对于维持森林生态系统的稳定性十分关键。

尽管目前已经开展了一些有关雌雄异株植物对干旱胁迫响应及机理的研究，但研究主要集中于叶片，如光合效能、用水效率、蒸腾速率和抗氧化防御系统等，而对植物的地下部分研究还未见报道。这可能是方法学的困难导致根的结构以及生物量积累等很难实现原位观察和测量。尽管如此，根特性仍然可能在性相关的生理差异中发挥关键作用。为了准确预测每个性别在不同微生境下的生长、存活和繁殖，亟待补充更多的有关性差异的地上和地下部分生理学信息。

嫁接指的是把一个植物组织插进另一个植物组织，使两个维管组织融合在一起成为一个完整的植株。被挑选作为根或将来发育成根的植物称为砧木，被挑选作为地上部分的植物称为接穗。通过嫁接可以巧妙地把植物的地上部分和地下部分连接在一起，可作为一种手段来区分植物地上部分和地下部分在一些生理学过程中发挥的不同作用。因此，通过对雌雄植株地上部分和地下部分的交互嫁接，可以确定雌雄植株对环境胁迫的差异响应是由根表达的特性和地上部分表达的特性决定的，抑或是由植株整体表达的特性决定的。此外，嫁接被广泛用于增强植物对非生物胁迫的抗性。通过嫁接有可能改善雌株的逆境容忍能力，对于改善雌株的存活、繁殖和提高雌雄性比将发挥重要作用。

§1.1 常用术语

本节将介绍雌雄异株、性二态性、交互嫁接、光合速率几个常用术语。

1.1.1 雌雄异株

雌雄异株（Dioecy）指在具有单性花的种子植物中，雌花与雄花分别生长于不同的株体。性别决定方式是 XY 型。仅有雌花的植株称为雌株，仅有雄花的植株称为雄株。有的植物雌株与雄株的染色体组成具有显著的差异。

1.1.2 性二态性

性二态性（Sexual dimorphism）是指同一物种的性别表现出不同的特征，特别是与生殖无关的特征。这种情况发生在大多数动物和一些植物中。差异可能包括次要性别特征、大小、体重（生物量）、颜色等。这些差异可能是细微的，也可能是巨大的，并且可能受到性选择和自然选择的影响。

1.1.3 嫁接

嫁接（Graft）是植物的人工繁殖方法之一，即把一株植物的枝或芽嫁接到另一株植物的茎或根上，使接在一起的两个部分长成一个完整的植株。嫁接的方式分为枝接、芽接和根接。嫁接是利用植物受伤后具有愈伤的机能来进行的。

1.1.4 光合能力

光合能力（Photosynthetic capacity，A_{max}）是衡量光合作用期间叶片能够固定碳的最大速率的指标。它通常指每平方米每秒固定的二氧化碳数量（$\mu mol\ CO_2 \cdot m^{-2} \cdot s^{-1}$）。光合能力受到羧化能力和电子传递能力的限制。

§1.2 雌雄异株植物的性二态性研究进展

雌雄异株植物是研究物种适应性的优良材料，尽管雌雄植株在第二性征和繁殖生态学方面存在差异，但它们在表型和生物学方面却有很多相似的地方。因此，研究雌雄异株植物形态学和生理学的性二态性，并探讨其功能意义是完全可行的。相比于花特性、生活史以及生物交互作用的研究，针对雌雄异株植物的生理学和营养器官形态学方面的研究还很少。

1.2.1 生殖成本假说

生殖成本假说认为，植物资源分配或开发是有限的，因而生殖成本的增加会导致适合度下降（繁殖能力和存活率下降）。在生理学水平，营养生长和生殖之间的资源分配存在此消彼长

(Trade-off) 的关系。植物可以通过减少营养生长或生殖,来抵消生殖的生理学成本。对于雌雄异株的木本植物来说,雌雄植株生殖成本的消耗时间和程度均不相同。相比于雄株,雌株消耗的净总成本会更多。性别间生殖分配的差异导致了雌雄生活史特性的相对差异。

1.2.2 生殖成本驱动下的雌雄异株植物的性二态性

由于雌株生殖资源成本比雄株高,因此雌株的选择压力更大,对资源的获取更多,而增加资源获取的生理学和形态学机制取决于哪一种资源对生殖的影响最大。例如,当碳固定被干旱限制时,雌株由于生殖成本较高,可能面临的选择压力更大,其光合生理和气孔行为、水压传导特性和植物组织水关系特征以及根系生长等将改变,以应对干旱对生殖的影响;而在营养贫瘠的环境中,雌株由于生殖成本更高,可能会面临选择压力,从而增加同化物向根的分配或增加根的营养吸收速率。总之,在胁迫条件下,由于生殖成本的差异,性相关生理学差异会更加明显,承担高生殖投入的性别对生理学胁迫将更加敏感,尤其在花期和果期。

§1.3 植物对干旱和铅胁迫的生理生态响应及研究进展

1.3.1 植物对干旱胁迫的生理生态响应及研究进展

由于全球气候的改变,水资源短缺逐渐成为限制森林生态系统生产力的重要因素。干旱频率的增加对森林生态系统树木的存活有很大的影响，导致森林生态系统生产力急剧下降。在中国，干旱、半干旱地区的总面积占全国总面积的一大半，而该区域的森林生态系统构成了我国陆地生态系统生物量、净初级生产力和生物多样性的相当大部分，干旱将对我国森林生态系统产生严重的不利影响。其中雌雄异株森林生态系统尤其脆弱，受气候变化的影响也更加明显。在半干旱地区，全球气候改变将引起温度梯度和降雨模式的明显改变，并增加极端气候事件的频率和强度（如干旱）。鉴于全球和区域性变化，了解森林生态系统对干旱的响应对于预测自然森林生态系统种群的变化和调整林业经营管理模式至关重要。由于所有生物地球化学过程均由气候驱动，干旱事件的增加不仅会影响森林生态系统的碳收益和碳损失，也会通过影响物候及树木的生物化学、生理学和形态解剖学，进而影响其生长发育。植物个体在受到干旱胁迫时，体内会产生一系列的形态、生理、生化、细胞及基因表达的变化，以减轻或延缓缺水对植物的伤害，这些变化通常体现在叶面积减少、叶片脱落、根冠比增加、气孔关闭、光合作用抑制、细胞渗透调节物增加、保护酶系统建立、水流阻力增加、叶片蜡质沉积、叶片能量耗散改变、景天酸代谢途径（CAM）启动（一些植物）、基因表达改变以及信号转导通路改变等方面。

1.3.1.1 干旱对植物光合作用的影响

干旱对光合作用的影响是植物对干旱胁迫生理学响应的一个最重要部分。干旱条件下，由于气孔关闭，从大气到羧化作用位点的 CO_2 扩散下降，导致光合作用下降。一些研究表明，干旱使得叶片含水量下降以及离子浓度增加，进而导致新陈代谢受损，而相比于气孔关闭，这种代谢损伤对光合作用的影响更大。此外，值得注意的是，叶片内部 CO_2 扩散 g_m 下降（如叶肉导度降低）也是干旱条件下光合作用损伤的一个潜在原因。

1. 扩散限制和生物化学限制

在植物的干旱胁迫响应中，气孔关闭是最早出现的一个生理学事件，由于气孔导度（g_s）与净光合作用（P_n）之间存在很强的正相关性，通常认为在干旱条件下，气孔关闭减少了 CO_2 吸收，从而导致光合作用下降。一些研究表明，在干旱条件下，如果增加 CO_2 浓度或剥掉叶片的上表皮，光合作用可以被完全恢复。该研究支持了气孔限制这一理论。但是另外一些研究却表明，通过这两种办法并不能将光合作用恢复到对照水平。这些相反的研究结论使得在干旱条件下，气孔或非气孔限制（生物化学限制）是否为光合作用下降的主要原因这一科学问题更具争议性。Tezara 等（1999）的研究对这一争议的影响很大，他们的研究表明，干旱条件下，叶绿体 ATPase 活性下降导致光合磷酸化受损是限制光合作用的主要因素。最近一些研究重新分析了基于气孔导度的数据，通过对这些数据的分析，已经就不同干旱胁迫强度对光合作用相关参数的影响这一问题达成了广泛共识。该共识具体内容如下。

阶段 1：轻度水分胁迫（$g_s > 0.15$ mol $H_2O \cdot m^{-2} \cdot s^{-1}$）。随着水分亏缺逐渐增加，$g_s$ 从最大值下降到约 0.15 mol $H_2O \cdot m^{-2} \cdot s^{-1}$，此阶段 g_s 下降是 P_n 降低的唯一原因。线性电子传递速率（ETR）、叶肉导度（g_m）、光合作用相关酶活性、光系统Ⅱ最大光量子效率（F_v/F_m）及最大羧化速率（$V_{c,max}$）等保持不变。随着胞间 CO_2 浓度（C_i）和羧化部位的 CO_2 浓度（C_c）下降，光呼吸速率逐渐增加。

阶段 2：中度水分胁迫（0.15 mol $H_2O \cdot m^{-2} \cdot s^{-1} > g_s >$ 0.05 mol $H_2O \cdot m^{-2} \cdot s^{-1}$）。在这一阶段，$g_s$ 进一步降低，并伴随 g_m 的大幅度下降。一些研究表明，g_m 降低与一些水通道蛋白丰度的改变有关，尽管如此，仍需要开展进一步的研究以充分阐

明 CO_2 的内部扩散是如何被调控的。在该阶段，ETR 出现了小幅度但明显的减少，同时，叶绿素荧光非光化学淬灭（NPQ）、抗氧化酶活性和非酶抗氧化剂数量细微增加。这些响应表明，叶子正在为更加严重的干旱胁迫做准备。在该阶段，如果使用传统的气体交换分析（即 P_n—C_i 曲线）来评估非气孔限制的存在，会得出 $V_{c,max}$ 下降这一错误结论。事实上，$V_{c,max}$ 的表面下降只是受到 g_m 降低的影响。P_n—C_c 曲线分析显示在该阶段 $V_{c,max}$ 几乎恒定不变。如果不考虑 g_m 的影响，还会错误估计 ETR 最大速率（J_{max}）。

阶段 3：严重水分胁迫（$g_s < 0.05$ mol $H_2O \cdot m^{-2} \cdot s^{-1}$）。当 g_s 低于该阈值时，尽管一小部分研究表明植物光合作用能力不受影响，但是大多数研究显示植物光合作用能力被严重破坏。有趣的是，在新陈代谢损伤的同时，通常伴随光合作用相关酶活性的抑制、叶绿素和蛋白质含量的下降以及光合作用系统的永久损伤，这些变化表明叶片正在经受氧化胁迫、衰老的诱发以及叶片养分的转运。新陈代谢损伤受调控系统的严格调控，包括较低的 g_s 下，C_c 降低而导致的 Rubisco 酶活性下降。但是，这些损伤是否来源于 Rubisco 激活状态的减少、酶浓度的减少、和/或酶抑制剂浓度的增加，则取决于植物类型和干旱胁迫处理方法。

2. 次级氧化胁迫

干旱条件下，当叶绿体中可利用的 CO_2 浓度急剧减少时，在卡尔文循环中，用于 CO_2 同化的电子利用减少，过多的电子可能被用于其他过程，如原初反应或热耗散。后者是胁迫条件下植物光保护的一个主要手段。但是，当这些过程饱和时，ETR 组件就会被过度还原，将电子转移给光系统 I 的氧分子或者进行米勒反应（Mehler reaction）。在这一过程中，活性氧（ROS），如超氧阴离子（O_2^-）、过氧化氢（H_2O_2）和羟基自由基大量产生，如果不被植物有效清除，就会对光合作用器官产生次级氧化损伤。

1.3.1.2 干旱对植物根部的影响

1. 干旱对植物根形态和解剖特性的影响

干旱胁迫对植物最常见的影响就是诱发根茎比增加。这种改变可能仅仅是地上部分生长减少，而根系生长并未发生变化；或者植物对根系生长的投资加大；抑或地上部分生长受抑制较多，而根生长受抑制较少。干旱条件下根茎比增加，植物可以拓展更大的土壤空间，并从中吸收更多的水分。然而在根茎比和干旱容忍性之间并未发现明显的联系。这种相关性的缺乏可能与植物根吸水能力存在差异有关。另外，在干旱条件下，根长度或根长密度（RLD）与干旱容忍性也密切相关。如 Grzesiak 等（1999）发现在干旱条件下，干旱容忍的玉米品种比干旱敏感的玉米品种具有更长的次生根。Jongrungklang 等（2011）观察到深层土壤的根密度和产量之间存在一定的相关性。然而值得注意的是，尽管在干旱条件下，根生长和干旱容忍性间可能存在相关性，但很明显，单独的根生长参数无法解释根的较高的水分吸收能力。Kondo 等（2000）发现，在严重干旱条件下，相比于水稻，玉米可以从土壤深层吸收更多的水，这不仅因为玉米根系统发达，而且因为玉米单位根表面积具有较高的水分吸收能力。因此，根显然存在某种内在的属性（如凯氏带），使得在干旱条件下，某些根具有更高的水分吸收效率。如研究发现，干旱条件下，植物根部凯氏带等障碍增加，从而降低了根的水分吸收能力（水压传导，L_p）。但最近研究也发现，这些凯氏带等障碍并不总是会减少根的水压传导，如 Ranathunge 等（2011）研究发现，凯氏带的发育、软木脂层的增加及厚壁组织的木质化并没有对根水压传导产生影响。

2. 干旱对植物根水压传导的影响

根水压传导特性对干旱非常敏感，很多研究发现，干旱胁迫会减少根水压传导。但是在一些特殊的环境条件下或对于特定的基因型，干旱则增加了根水压传导。如 Wikbergi 和 Ogreni（2007）对柳树的研究发现，在中度干旱条件下，地上部分水压传导和根水压传导之间的比值越高，生物量的积累就越多。然而，McLean 等（2011）发现，当树木一半根生长在干燥区（几乎没有水分），另一半根生长在湿润区时，生长在湿润区的根可以通过增加水压传导和 PIP_1 水通道蛋白的数量来提高根吸水能力。该研究表明，水通道蛋白对于水压传导的调控至关重要，并且在干燥根和湿润根之间存在信号交流。

3. 干旱对植物水通道蛋白的影响

干旱条件下，根可以通过调节水通道蛋白（PIPs 蛋白）的活性而调节水分吸收。一些研究发现，干旱条件下，水压传导和 PIPs 蛋白丰度之间存在一定相关性；而在另外一些研究中，这种相关性又消失。过表达或敲除特定的水通道蛋白基因可以作为一种手段，来研究干旱条件下水通道蛋白如何调节水压传导。如 Lian 等（2004）发现在干旱敏感水稻基因型中，过表达 PIP_1 基因可以提高其在干旱条件下的水压传导，并且与非转基因的干旱敏感水稻基因型相比，维持了较好的水分状态。

1.3.2 植物对铅胁迫的生理生态响应及研究进展

随着生态环境的恶化，植物受到越来越多污染物的影响。这些污染物主要通过土壤或大气进入植物系统。在影响植物的常见污染物中，铅是毒性最强且最常见的污染物之一。铅被广泛用于各种工业过程，在环境（土壤、水、大气和生物体）中频繁出现。据统计，2009年采矿业可回收铅的生产在中国、澳大利亚和美国分别为1690万吨、516万吨和400万吨。人类使用铅的历史很悠久，目前并未发现铅在生物体中有任何已知的生物学功能。另外，基于出现的频率、毒性和人类接触的可能性，铅仅次于砷，是第二大剧毒物质。因此，铅在植物体内的转移规律被大量研究，尤其是在食品质量安全、植物修复技术开发和生物监测等领域。已知铅对生物体产生形态学、生理学和生物化学等多种有害影响，包括对植物生长发育、根伸长、种子萌发、幼苗发育、蒸腾作用、叶绿素生产、叶绿体片层结构和细胞分裂等的影响。植物已经开发出各种方法来应对有毒金属的影响。这些内部的脱毒机制主要包括金属选择性吸收、排泄、特定配体的络合作用以及重金属离子的区隔化。植物对铅暴露的各种响应常常被用作环境质量评估工具（生物指示）。而对于生态毒理学的开发来说，了解重金属在植物中的吸收、转移及毒性机理也是非常必要的。

1.3.2.1 铅对植物生长的影响

当植物暴露于铅，即使是微摩尔的水平，也会对一些植物的成长产生负面影响。铅对幼苗的生长和发育有强烈的抑制作用。较低铅浓度下，植物的根和地上部分的生长均受到抑制，尤其是植物的根，这可能与根部积累大量的铅有关。铅诱发植物的根肿胀、弯曲及变短变粗，并且会增加单位根长的二级根数量。Jiang 和 Liu（2010）发现，经48～72 h铅处理后，大蒜根细胞出现了线粒体肿胀、嵴丢失、内质网和高尔基体空泡化以及质膜损伤等症状。Arias 等（2010）也发现，在豆科灌木中，铅明显抑制了根的伸长。另外，高剂量铅也会影响植物生物量的累积。在严重铅胁迫下，植物表现出明显的生长抑制症状，如叶片变小、叶片数量减少、叶片易碎以及叶片远轴面（下表面）呈现暗紫色。铅对植物生长的抑制可能归因于营养代谢障碍和光合作用损伤。大多数情况下，铅对植物生长的毒害作用取决于铅胁迫时间和铅胁迫剂量。而在较低的铅浓度下，这种影响并不明确，生长抑制与生物量的减少之间没有必然联系。此外，铅毒性产生的影响也随着植物的种类而变化，超积累植物比敏感植物更耐铅毒。

1.3.2.2 铅对植物光合作用的影响

光合作用下降是铅毒的一个明显症状。铅对光合作用的抑制是由于铅的间接影响而不是直接影响。这些间接影响包括以下方面：①铅对蛋白质 N 和 S 配体的亲和，叶绿体超微结构的扭曲和变形；②叶绿素合成抑制开始阶段，铁氧化还原蛋白 $NADP^+$ 还原酶和 δ-氨基乙酰丙酸脱水酶（ALAD）的活性降低；③质体醌和类胡萝卜素合成的抑制；④电子传递系统障碍；⑤气孔关闭，导致 CO_2 浓度下降；⑥必需元素（如 Mn 和 Fe）吸收阻碍以及二价阳离子的铅置换；⑦卡尔文循环酶活性降低；⑧叶绿素酶活性增强。

1.3.2.3 铅对植物水分状态的影响

很多研究已经发现铅会破坏植物的水分状态。如铅胁迫下，植物蒸腾作用降低，组织含水量下降。蒸腾作用降低可能是叶表面积减少，然而，一些植物具有较高的气孔密度，因而有能力应对这些影响。铅也降低了植物细胞壁的可塑性，从而对细胞膨压产生影响。另外，糖和氨基酸等分子浓度的降低也影响了细胞肿胀，从而进一步加剧了铅对细胞膨压的影响。膨压的变化尤其是保卫细胞膨压的变化，对气孔的开关产生影响。为了维持细胞膨压，在铅胁迫条件下，植物合成大量的渗透物质，如脯氨酸。

Pb^{2+}的存在还会导致脱落酸（ABA）在植物根部和地上部分大量累积，致使气孔关闭，而气孔关闭大大限制了植物的气体交换和蒸腾作用的水损失。此外，Elibieta 和 Miroslawa（2005）的研究表明，由于铅沉积到植物叶片的表皮层，导致植物叶片的呼吸作用也受到影响，而氧化磷酸化障碍和呼吸障碍使得 CO_2/O_2 失衡，进而破坏了植物的水分状态。

1.3.3 植物对干旱和重金属交互胁迫的生理生态响应及研究进展

迄今为止，植物对单一胁迫响应的形态学特征和生理学特性的研究已经大量展开，而植物对多种环境胁迫响应的研究则较少。研究发现，植物对干旱和重金属胁迫的响应存在一定的相似性，如干旱和重金属都减少了植物细胞的延长。植物对干旱和重金属胁迫的响应也存在一定差异，如在干旱条件下，植物会相对增强根的生长（相比于叶片生长），以便从土壤中获得更多的水分；重金属胁迫则抑制了根的生长。而且，植物暴露在重金属环境下可能会影响其应对干旱胁迫的能力。重金属对干旱容忍性影响的大小取决于重金属的化学特性、重金属的生物可利用性以及重金属对植物形态功能特性诱导响应的类型和程度等。例如，重金属诱导的植物根改变、木质部解剖学结构改变、必需营养供应减少、光合作用速率和植物生长减少等会进一步影响植物的水分吸收和运输能力。

§1.4 嫁接植物对非生物胁迫的响应

环境胁迫，如水、温度、营养、光、氧的可用性、金属离子浓度和病原体，在很大程度上限制了植物的生产力和植物的种群分布。一种特殊的方法就是把对环境胁迫敏感的植物嫁接到环境胁迫抗性较强的砧木上，从而提高敏感植物对环境胁迫的抵抗能力。目前，嫁接已经被广泛用于减少土壤携带病原菌对植物的侵染以及增强植物对非生物胁迫的抗性，如盐胁迫、碱胁迫、营养缺乏、重金属、热胁迫、水分胁迫及有机物污染等。

1.4.1 嫁接植物对水分胁迫的生理生态响应及研究进展

1.4.1.1 嫁接植物对干旱胁迫的生理生态响应及研究进展

干旱、半干旱地区，水资源缺乏一直是制约植物生长的关键因素。一种提高干旱条件下植物生产力和水分使用效率的方法就是将高产基因型的植物嫁接到耐旱性强的植物上。Sanders 和 Markhart（1992）的研究发现，干旱条件下，嫁接菜豆（*Phaseolus vulgaris* L.）接穗的渗透势是由砧木决定的，而水分充足条件下，其接穗的渗透势则取决于地上部分。Serraj 和 Sinclair（1996）的试验则发现，砧木或接穗提高了嫁接大豆（*Glycine max* L.）的干旱容忍性，并提高了其固氮能力。Clearwater 等（2004）开展的嫁接干旱试验则证明，耐旱性强的砧木可以大大提高接穗的抗旱性。对比之下，Abadelhafeez 等（1975）的研究发现，将根系吸水能力强的茄子作砧木，吸水能力差的番茄作接穗，并没有提高番茄的耐旱性；而把小西瓜嫁接到一种耐旱性强的西瓜砧木上，在干旱条件下，其产量比未嫁接的提高了 60%，这可能是因为耐旱砧木有较强的水分吸收能力、营养吸收能力（表现为叶组织含有较高的 N、K 及 Mg 浓度）以及较高的 CO_2 同化能力。

1.4.1.2 嫁接植物对水淹胁迫的生理生态响应及研究进展

水淹是非常重要的非生物胁迫，严重影响了水淹敏感植物的生长和生物量累积。由于在水淹条件下气体扩散缓慢，加之微生物和植物根又对氧气进行消耗，导致植物缺氧。水淹引起的这些问题可以通过种植水淹容忍的植物或嫁接水淹敏感的植物到水淹容忍的植物上解决。例如，把水淹敏感的苦瓜嫁接到水淹容忍的丝瓜上，明显改善了苦瓜的水淹容忍性。光合作用速率、气孔导度、蒸腾、可溶性蛋白质和/或 Rubisco 活性降低的快慢可能是植物具有不同的水淹容忍性的原因。研究发现，将水淹敏感的黄瓜接穗嫁接到水淹容忍的南瓜砧木上，明显改善了黄瓜叶片在水淹条件下叶绿素含量下降这一问题，这可能与木质部树液中的一种化学信号刺激了地上部分乙烯的生物合成有关。研究还发现，水淹条件下嫁接西瓜形成了不定根和通气组织，而非嫁接西瓜则未出现这些结构。

1.4.2 嫁接植物对温度胁迫的生理生态响应及研究进展

1.4.2.1 寒冷胁迫

温度是最重要的环境因素之一，严重影响了植物的生长和发育。在冬天，甚至春天和秋天，植物常常面临低温胁迫，进而导致植物生产力下降甚至死亡。植物种类不同，其低温胁迫阈值也不同，对低温敏感的果蔬类，其低温胁迫阈值为 $8℃\sim12℃$。在 20 世纪 60 年代末期，Hori 等（1970）对黄瓜开展了嫁接试验以提高其对低温的耐受性，但是结果表明，嫁接并没有改善其低温耐受性。在葫芦科和茄科开展的各种接穗－砧木嫁接组合试验也没有提高嫁接材料的低温耐受性，在低温条件下，嫁接植物的生长状况和产量均没有提高。在温室条件下，番茄的嫁接植株比未嫁接植株产果率更高，但在野外低温条件下，这种优良性状又几乎消失殆尽。Zijlstra 和 den Nijs（1987）证明，在较低的白天和夜晚温度（18℃/7℃）下，番茄嫁接植株幼苗开花期提前，果实产量提高，这主要是由于挑选了一个对低温有明显耐受性的砧木。然而，由于在驯化过程中存在近亲繁殖，导致嫁接植株遗传多样性小，对低温变化的适应能力也小，而相比之下，近缘野生种的低温容忍性变异则较大，尤其是在昼夜温差很大的高海拔地区。

1. 耐低温砧木的低温耐受性机理

低温会影响根系生长、大小、结构和功能。在低温条件下，根（砧木）调控地上部分（接穗）的生长主要归因于水的黏度、根压、水压传导、代谢活性、植物激素合成和向上运输以及根的营养吸收能力。砧木能在多大程度上缓解接穗对低温胁迫的敏感性主要取决于砧木的生长状况和功能状况、砧木和接穗之间的相互作用、嫁接组合的生理年龄以及低温胁迫的持续时间和强度。

（1）根的生长和结构。

由于根的隐蔽性，有关温度对根的表型影响研究较少。根成像技术将帮助我们获得较多的关于温度改变如何影响根结构的相关信息。在低温条件下，相比于低温敏感的砧木，耐低温的砧木维持了较高的根生长速率。研究发现，将番茄嫁接到低温容忍的砧木上，可提高其根茎比在低温条件下的调节能力。尽管这种从叶（源）到根的同化物分配转换的生理学机制仍不清楚，但是这种适应性响应恢复了根茎之间的功能平衡，使根克服了水分和养分的吸收限制。此外研究发现，耐低温番茄砧木的迅速生长似乎与一级侧根的迅速生长有关。但是，Lee 等（2004a）的研究却发现，低温敏感的黄瓜和低温容忍的黑子南瓜的根系结构并没有明显差异。

（2）营养吸收。

值得注意的是，有关嫁接植株在低温胁迫下的营养吸收和转运的研究资料很少。目前开展的相关研究主要集中在葫芦科，研究显示，在低温胁迫下，相比于低温敏感种，低温容忍种的大量元素，尤其是硝酸盐和磷酸盐的吸收和转运均有所增加；微量元素 Mn、Cu 和 Zn 的含量增加，而 Fe 通常不受低温影响（Li & Yu，2007）。对于绝大多数营养元素，每单位面积根的营养吸收率下降，可能归因于低温条件下砧木根面积的增加。在低温条件下，番茄的大量营养元素 P 的吸收明显被抑制，而有效 P_i 的下降可能限制了其光合作用。目前没有数据显示，在番茄嫁接苗的低温胁迫研究中，耐低温的砧木能够减缓地上部分（接穗）有效磷的下降。一般来说，在低温胁迫下，营养元素吸收受低温影响小的根系（砧木）通常有较高的新陈代谢活性、呼吸速率以及 H^+-ATPase 质子泵活性。

（3）水的吸收、转移及渗透调节。

低温对根影响的研究主要集中在水分吸收方面。在低温条件下，耐低温的砧木通过增加根的水压传导，减少细胞壁凯氏带的感应、脂质过氧化反应和气孔关闭，能够克服低温对水分吸收的限制。然而，Bloom 等（2004）的研究表明，低温条件下根对水分的吸收并不是由于较高的水压传导能力，而是因为分配到根的生物量增加，因此，弥补了低温条件引起的水分移动能力的下降。相比于低温敏感的植株，在低温条件下，耐低温植株根的水压传导能力下降较少，所以水的纵向运输受到的影响较小。当把不耐低温的黄瓜嫁接到耐低温的黑子南瓜上时，在低温条件下，黄瓜叶片的含水量下降较少，渗透调节物质，如氨基酸、季铵化合物、多元醇和糖类大量累积，表明其新陈代谢并未受到干扰。

（4）脂质过氧化和抗氧化物质。

低温胁迫使得低温敏感植物的根部产生了大量的 ROS，例如过氧化物、过氧化氢以及羟基自由基，引发了不饱和膜脂的过氧化反应，进而导致膜的硬度增加，电解质、水和可溶性物质泄露出根细胞。通过测量丙二醛浓度很明显地发现，低温敏感的黄瓜嫁接到低温容忍的砧木上能够减少低温胁迫下黄瓜脂质过氧化和电解质泄露程度。而比较冷敏感和冷容忍的黄瓜砧木发现，二者冷敏感性的差异与 ROS 的累积差异直接相关。对番茄来说，嫁接几乎能够完全抑制低温所诱导的叶片 H_2O_2 的累积。在低温条件下，随着 ROS 的增加，脱毒的抗氧化物质也明显增加。然而，许多关于低温胁迫下嫁接植物的抗氧化物浓度和 ROS 清除酶活性的研究结果并不一致，这似乎取决于实际处理温度，以及是整株植物暴露于低温，还是只有砧木暴露于低温。此外研究发现，将低温敏感的接穗嫁接到低温容忍的砧木上，ABA 和细胞分裂素（这两个激素主要在根部合成）在植物体内的运输能力提高，由于 ABA 和细胞分裂素能够上调 ROS 清除系统，这也可能是嫁接植株 ROS 清除酶活性提高的原因之一。

（5）库源关系。

在低温条件下，新陈代谢和生长能力的抑制有助于根库强度的下降，例如，根从韧皮部获取光合作用产物的能力下降，使得碳水化合物从源叶的输出减少，增加了其在叶片中的累积，进而导致比叶面积减少。而比叶面积的减少与细胞大小的横向增加会导致叶片厚度增加，致使吸光面积减少，因此阻碍了光合作用和生物量的累积。另外，叶片中碳水化合物的累积也可以下调光合作用相关基因的表达，进而抑制光合作用。因此，温度胁迫对砧木库强的抑制程度是影响叶片形态、光合作用以及地上部分生长能力的重要因素。事实上，对很多物种进行的嫁接研究表明，在低温条件下，耐低温的砧木能提高植物的光合作用。除了较高的库容量，在低温条件下，低温容忍的砧木也可以给接穗提供更多的水、营养和激素，而这些因素均会影响光合作用。此外，耐低温的砧木对光合作用的促进也可能是叶绿体中存在较少的 ROS，进而减少了对 Rubisco 和 D_1 蛋白的降解。在非生物胁迫下，砧木生长（库强）的维持似乎与胁迫耐性相关。目前，根－地上部分信号传导的生理学知识还很缺乏。研究表明，除了植物激素，一些特定的 RNA 分子作为一种遗传信息，通过在韧皮部中的传导，可以协调砧木和接穗的生长和发育。

（6）植物激素。

受低温的诱导，植物根部合成激素的变化不仅对根的生长（库强）产生影响，而且对根与地上部分的激素信号传导产生影响，进而会改变地上部分的生理机能及生产力。目前，这方面的研究相当缺乏。研究表明在低温条件下，地上部分生物量的大幅度下降，是因为水流动变慢导致根部合成的植物激素（如细胞分裂素和赤霉素）向上运输变慢，而不是因为激素的生物合成速率变慢。低温对番茄地上部分生长的抑制（如限制了叶片的扩张和茎的生长）可以通过在根尖或根的生长介质中添加少量的赤霉素而部分抵消。但是，添加激动素或苄基氨基嘌呤（BAP，一种人工合成的细胞分裂素）只略微改善了低温条件下番茄地上部分的生长状况。低温除了降低了细胞分裂素和赤霉素的向上运输，还增加了 ABA 的向上运输，并诱导其在地上部分累积。目前大量证据表明，根部的 ABA 可以影响根细胞膜部位的水通道蛋白的构象，从而更有利于水分的吸收。而在地上部分，ABA 则通过调节气孔的张开而控制蒸腾速率。最近的实验提供了令人信服的证据，表明 ABA 对于维持根和地上部分的生长是必需的，尤其是叶片的扩张，与先前通常认为的 ABA 抑制叶片扩张的理论恰好相反。其机制一部分是对乙烯合成的抑制，另一部分则是一种不依赖于乙烯的机制。在低温胁迫下乙烯（ACC）合成增加。

此外，研究表明低温胁迫下地上部分生长的下降与吲哚乙酸（IAA）的累积有关，这主要是在低温条件下，IAA 从地上部分向根部的运输下降及细胞分裂素氧化酶活性增加，致使地上部分的细胞分裂素浓度下降。黄瓜的嫁接试验发现，在低温条件下对低温越敏感的砧木，其地上部分木质部树液中的 ABA 含量增加越多并且增加速度越快，并伴随有细胞分裂素的减少。与该研究相反，低温刺激了黑子南瓜细胞分裂素的合成，导致其根木质部树液中细胞分裂素浓度增加。

1.4.2.2 高温胁迫

高温胁迫诱导植物产生了一系列复杂的形态、生理、生化和分子的变化，进而影响了植物的生长和生产力。这些有害的影响包括生长减缓、光合作用速率下降、呼吸作用增加、同化物向果实的分配增加、渗透调节损伤、氧化损伤、水分吸收下降、离子吸收下降以及细胞脱水等。另外，高温激活了植物的胁迫响应机制，例如蛋白质合成（如热休克蛋白）的变化、解毒能力的提高、渗透保护的增强以及酶和膜系统稳定性的提高。而砧木能否改善这些防御过程，最终提高整个植物的耐热性，还需开展进一步的研究。

1. 砧木作为一种工具增加了植物的高温容忍性

早期的一些嫁接试验并没有改善植物的耐热性，后来的研究发现使用茄科的植物作为砧木，可以成功地提高植物的耐热性，如使用番茄作为砧木可以授予植物一定程度的耐热性。由于茄子能很好地适应炎热干燥气候以及忍受较高的土壤温度，将番茄嫁接到茄子砧木上可以提高番茄的高温容忍性。

2. 砧木耐高温的机制

与低温相比,高温严重减少了根的伸长并增加了平均根直径。乙烯合成的增加似乎在高温胁迫响应中发挥了关键作用。乙烯生物合成抑制剂可以部分缓解高温胁迫对根伸长生长和横向生长的影响,对叶片水状态和气孔打开有积极的影响,而光合作用能力和生物量累积并没有得到改善。高温条件下,光合作用能力和生物量累积的下降可能受营养亏缺介导的气孔限制影响。高温也可以引起根部和地上部分多种矿质营养元素的缺乏(如 P 和 Fe),而矿质营养元素缺乏能增加乙烯的生产。另外,高温还能减少膜的稳定性,致使植物新陈代谢紊乱。研究发现,将番茄嫁接到一个耐热的砧木上,其过氧化氢浓度下降,表明番茄受到的氧化胁迫降低。还有研究表明,在高温胁迫下将番茄嫁接到茄子砧木上,可以减少其电解质泄露,说明其膜损伤减少,溶质和水的保持能力增强。对比自嫁接的番茄,番茄嫁接到茄子砧木上也表现出较低的脯氨酸水平,但抗坏血酸浓度明显增加。

试验方法篇

第 2 章　材料与试验处理方法

§2.1　植物材料的采集和培养

2.1.1　植物材料的采集

本试验以青杨（*Populus cathayana*）雌雄植株为材料。在青海省（东经 101°35′，北纬 35°56′，海拔 2450 m，年均温 3.7℃，1 月月均温 −11.2℃，7 月月均温 16.0℃，年降雨量 350 mm）河谷两侧的坡地选取 15 个种群，每个种群各选择 2 个雌雄个体，总共采集 60 个个体，包括 30 个雌树和 30 个雄树。为了避免个体间的无性繁殖，确保种群内个体间的基因型差异，种群内个体间的采样距离至少保持 15 m。

2.1.2　植物材料的培养

青杨雌雄枝条于 2010 年 3 月下旬扦插于中国科学院成都生物研究所温室内，该温室日温变化幅度为 19℃～28℃，夜温变化幅度为 12℃～18℃，相对湿度变化幅度为 40%～85%。约两个月后，幼苗长势稳定，此时根据幼苗的基径和高度，选择大小一致的雌雄健康幼苗各 240 株，将其移栽到含 32 kg 匀质土壤（pH=7.1±0.1）的塑料盆中，每盆移栽 1 株幼苗。另外，在处理过程中，定期交换栽植盆位置，避免由于方位不同对植物产生影响。1 年之后（2011 年 3 月），根据幼苗的基径和高度，选择大小一致的雌雄健康一年生幼苗各 80 株用于嫁接。嫁接 8 周之后，水分胁迫处理 8 周。水分梯度的选择根据取样地的主要气候条件和土壤水分状况而确定。另外在处理过程中，定期交换栽植盆位置，避免由于方位不同对植物产生影响。

为了确保嫁接的成功率和存活率，由两个有经验的嫁接工人指导嫁接。本书试验总共构建了 4 种嫁接组合：雄性接穗＋雄性砧木（M/M）、雌性接穗＋雌性砧木（F/F）、雄性接穗＋雌性砧木（M/F）、雌性接穗＋雄性砧木（F/M）。

§2.2 试验处理方法

2.2.1 铅和干旱试验处理方法

试验采用 3 因素的完全随机设计：2 性别（雄、雌）× 2 水分梯度（最大田间持水量的 100%、50%）× 2 个硝酸铅浓度 [0 mg·kg^{-1}、900 mg·kg^{-1} Pb(NO$_3$)$_2$ 添加进土壤或 0 mg、600 mg Pb(NO$_3$)$_2$ 添加到叶面]。具体的处理过程为：在铅添加到土壤的处理（Pb$_{soil}$）中，处理开始的最初 15 天将溶解有 1.92 g Pb(NO$_3$)$_2$ 的 250 mL 去离子水均匀喷洒到土壤中，最终的铅含量为 900 mg·kg^{-1} Pb(NO$_3$)$_2$ 干土。在铅喷洒到叶面的处理（Pb$_{leaf}$）中，处理开始的最初 15 天用喷水壶将装有 500 mL 80 mg·L^{-1} Pb(NO$_3$)$_2$ 的溶液，从顶部均匀喷洒到每个植株，最终每个植株接受的 Pb(NO$_3$)$_2$ 为 600 mg。在 Pb$_{leaf}$ 处理中，为了避免污染土壤，我们在土壤上覆盖了一层塑料布。试验中铅浓度的选择参照铅矿附近土壤和雨水中的铅浓度。

在控制水分试验中,两个水分梯度的选择是根据铅矿附近普遍的气候条件和土壤水分状况而确定。在水分充足的处理中,每隔1天对盆进行称重,并复水到100%田间持水量(土壤含水量维持在41.9%)。在干旱胁迫的处理中,盆土维持50%的田间持水量(土壤含水量维持在20.5%)。在此期间,为了防止土壤水分的蒸发或渗漏,用塑料袋在每株茎基处将盆密封。另设10盆对照盆,用于测定土壤表面蒸发。每盆以木棒代替树苗栽在盆中,并用同样方法密封。无论处理和对照,均采用每天称重补水的方法控制盆中土壤水分含量。在苗木的快速生长期,利用植物鲜重(Y,g)和植株高度(X,cm)之间的关系式:$Y=0.975+0.112X$ ($R^2=0.968$,$P<0.001$)来校正随植株生物量的变化而引起的控水量误差。在整个试验过程中,每盆施加8 g缓释肥。处理开始于2011年5月10日,结束于2011年8月10日。

2.2.2 干旱和嫁接试验处理方法

试验采用完全随机设计:4种嫁接组合(M/M、F/F、M/F和F/M)×2个水分梯度(最大田间持水量的100%、30%)。每个水分梯度包括80个个体(每个嫁接组合20个个体)。在充分浇水的处理中,每隔1天对盆进行称重,并复水到100%田间持水量(土壤含水量维持在37.6%)。在干旱胁迫的处理中,土壤维持30%的田间持水量(土壤含水量为12.5%)。水分控制方法同铅和干旱试验。在整个试验过程中,每盆施加8 g缓释肥。处理开始于2011年5月6日,结束于2011年8月6日。

第3章 气体交换测量方法

§3.1 光合作用分析

3.1.1 光合作用相关色素分析

色素的提取方法参照 Lichtenthaler（1987）。在气体交换测量之后，立即将测量的叶子剪下，然后用打孔器（直径为 0.8 cm）从每张叶片的中部取得 3 个叶圆片，制备叶圆片时应尽量避开叶脉。将取好的叶圆片称重后立即浸入 80% 丙酮中避光浸提，直至其完全变白。用分光光度计（Unican UV-330, Unicam, Cambridge, UK）读取上清液在 470 nm、646 nm 和 663 nm 的吸收值。叶绿素含量根据 Porra 等（1989）描述的公式进行计算。总叶绿素含量（T_{Chl}）为叶绿素 a 和叶绿素 b 相加所得之和。

3.1.2 气体交换分析

在每个处理中，随机选择 5 个个体进行光合作用气体交换测定。使用 LI-6400 便携式光合作用测定系统（LI-COR Inc. Lincoln, Nebr.）选取第四叶位、健康及完整无损的叶片进行气体交换测定。测定时间选择在 2011 年 7 月晴朗上午的 08∶00—

11:30进行。通过预试验确定试验材料的饱和光量子密度（$PPFD$）。在测量之前，用 LI-6400 便携式光合作用测定系统的 LED 光源，在饱和的 $PPFD$ 下照射 10~30 min 以获得完全的光合作用诱导。具体测量条件如下：叶温 25℃，叶空气水汽压亏缺（VPD）（1.5±0.5）kPa，$PPFD$ 1500 $\mu mol·m^{-2}·s^{-1}$，相对空气湿度 50%，大气 CO_2 浓度（400±5）$\mu mol·mol^{-1}$。一旦气体交换达到稳态，就开始记录数据。在以下 $PPFD$ 测量净光合作用速率对 $PPFD$ 的响应（光响应曲线）：0 $\mu mol·m^{-2}·s^{-1}$、20 $\mu mol·m^{-2}·s^{-1}$、50 $\mu mol·m^{-2}·s^{-1}$、80 $\mu mol·m^{-2}·s^{-1}$、100 $\mu mol·m^{-2}·s^{-1}$、120 $\mu mol·m^{-2}·s^{-1}$、150 $\mu mol·m^{-2}·s^{-1}$、200 $\mu mol·m^{-2}·s^{-1}$、300 $\mu mol·m^{-2}·s^{-1}$、400 $\mu mol·m^{-2}·s^{-1}$、600 $\mu mol·m^{-2}·s^{-1}$、800 $\mu mol·m^{-2}·s^{-1}$、1000 $\mu mol·m^{-2}·s^{-1}$、1200 $\mu mol·m^{-2}·s^{-1}$、1400 $\mu mol·m^{-2}·s^{-1}$、1600 $\mu mol·m^{-2}·s^{-1}$、1800 $\mu mol·m^{-2}·s^{-1}$ 及 2000 $\mu mol·m^{-2}·s^{-1}$。用 Prioul 和 Chartier（1977）描述的非直角双曲线方程进行模型拟合。通过拟合数据到模型功能得到光饱和的光合作用速率（A_{max}），进一步通过对 0~200 $\mu mol·m^{-2}·s^{-1}$ $PPFD$ 内 P_n 与 $PPFD$ 的线性回归得到表观量子产量（AQE）和光补偿点（L_{cp}）。P_n 对 C_i 的响应曲线的制备方法如下：首先将叶子置于其生长环境的 CO_2 浓度（400 $\mu mol·mol^{-1}$）下，直到其气体交换达到稳态，然后将 CO_2 浓度按照 9 个梯度（400 $\mu mol·mol^{-1}$、300 $\mu mol·mol^{-1}$、200 $\mu mol·mol^{-1}$、150 $\mu mol·mol^{-1}$、120 $\mu mol·mol^{-1}$、100 $\mu mol·mol^{-1}$、80 $\mu mol·mol^{-1}$、50 $\mu mol·mol^{-1}$ 和 0 $\mu mol·mol^{-1}$）一直降到 0 $\mu mol·mol^{-1}$，之后再按照 7 个浓度梯度（400 $\mu mol·mol^{-1}$、500 $\mu mol·mol^{-1}$、600 $\mu mol·mol^{-1}$、800 $\mu mol·mol^{-1}$、1000 $\mu mol·mol^{-1}$、1200 $\mu mol·mol^{-1}$ 和 1500 $\mu mol·mol^{-1}$）将 CO_2 浓度逐渐增加到 1500 $\mu mol·mol^{-1}$。最后取 0~200 $\mu mol·mol^{-1}$ CO_2 区段绘

制 P_n 与 C_i 的直线回归曲线,用该回归曲线得到 CO_2 补偿点（Γ）和羧化效率（CE）。

§3.2 叶绿素荧光分析

3.2.1 叶绿素荧光主要参数

F_0：固定荧光,初始荧光（Minimal fluorescence）,也称基础荧光、0 水平荧光,是光系统Ⅱ（PSⅡ）反应中心处于完全开放时的荧光产量,它与叶片叶绿素浓度有关。

F_m：最大荧光产量（Maximal fluorescence）,是 PSⅡ反应中心处于完全关闭时的荧光产量。可反映经过 PSⅡ的电子传递情况。通常叶片经暗适应 20 min 后测得。

F：任意时间实际荧光产量（Actual fluorescence intensity at anytime）。

F_a：稳态荧光产量（Fluorescence instable state）。

F_m/F_0：反映经过 PSⅡ的电子传递情况。

$F_v = F_m - F_0$：可变荧光（Variable fluorescence）,反映了 QA 的还原情况。

F_v/F_m：PSⅡ最大光化学量子产量（Optimal/maximal photochemical efficiency of PSⅡ in the dark,Optimal/maximal quantum yield of PSⅡ）,反映 PSⅡ反应中心内禀光能转换效率（Intrinsic PSⅡ efficiency）或最大 PSⅡ的光能转换效率（Optimal/maximal PSⅡ efficiency）,叶片经暗适应 20 min 后测得。非胁迫条件下,该参数的变化极小,不受物种和生长条件的影响；胁迫条件下,该参数明显下降。

F'_v/F'_m：PSⅡ有效光化学量子产量（Photochemical efficiency of PSⅡ in the light），反映开放的 PSⅡ 反应中心原初光能捕获效率，叶片不经过暗适应在光下直接测得。

$(F'_m-F)/F'_m$ 或 $\Delta F/F'_m$：PSⅡ 实际光化学量子产量（Actual photochemical efficiency of PSⅡ in the light），它反映PSⅡ反应中心在有部分关闭情况下的实际原初光能捕获效率，叶片不经过暗适应在光下直接测得。

荧光淬灭分为两种，即光化学淬灭和非光化学淬灭。光化学淬灭以光化学淬灭系数代表：$q_P=(F'_m-F)/(F'_m-F'_0)$；非光化学淬灭有两种表示方法：$NPQ=F_m/F'_m-1$ 或 $q_N=1-(F'_m-F'_0)/(F_m-F_0)=1-F'_v/F_v$。

表观光合电子传递速率以 $[(F'_m-F)F'_m]\times PFD$ 表示，也可写成 $\Delta F/F'_m\times PFD\times 0.5\times 0.84$，其中，系数 0.5 是因为 1 个电子传递需要吸收 2 个量子，而且光合作用包括 2 个光系统；系数 0.84 表示在入射的光量子中被吸收的占 84%，PFD 是光子通量密度；表观热耗散速率以 $(1-F'_v/F'_m)\times PFD$ 表示。

F_{mr}：可恢复的最大荧光产量，它的获得是在荧光 P 峰和 M 峰后，当开放的 PSⅡ 最大荧光产量平稳时，关闭作用光得到 F'_0 后，把饱和光的闪光间隔期延长到 180 s/次，得到一组逐渐增大（对数增长）的最大荧光产量，将该组最大荧光产量放在半对数坐标系中即成直线，该直线在 Y 轴的截距即为 F_{mr}。以 $(F_m-F_{mr})/F_{mr}$ 可以反映不可逆的非光化学淬灭产率，即发生光抑制的可能程度。

3.2.2 叶绿素荧光分析方法

用于叶绿素荧光测量的叶片与用于光合作用测量的叶片为同一叶片。使用 PAM 叶绿素荧光仪（PAM 2100，Walz，Effeltrich，Germany），并根据 van Kooten 和 Snel（1990）的方法测量和计算叶绿素荧光动力学参数（F_v/F_m，PSⅡ光化学效率；$\Phi_{PSⅡ}$，PSⅡ有效光量子产量；q_N，非光化学淬灭系数；q_P，光化学淬灭系数）。具体测量方法为：首先将叶子用叶夹夹好，在暗处适应至少 30 min，然后测量最小荧光（F_0）和最大荧光（F_m），之后用 250 $\mu mol \cdot m^{-2} \cdot s^{-1}$ 的活化光（该光强为测量时温室的光密度）照射叶片，然后去除活化光，再用远红光照射 3 s，以测量最小荧光（F_0'）。最后用 8000 $\mu mol \cdot m^{-2} \cdot s^{-1}$ 的饱和白光照射 0.8 s，测量 F_m 和 F_m'。

第 4 章 生长参数及生理指标测量方法

§4.1 生长和形态参数测量

4.1.1 干重测定

试验结束后,收获生物量样本,并分成叶、茎和根,在70℃烘干约48 h 直至恒重并称量,然后计算干物质的累积。

4.1.2 叶肉细胞形态的透射电镜观察

透射电镜材料处理参照 Zhao 等（2009）的方法,并有一些改动。每个处理随机挑选 3 株植株,取每株植株第四叶位叶片（从上往下数）,将其用磷酸缓冲液冲洗干净,避开中脉在叶片中上部剪取 2 mm×2 cm（宽×长）的长条形叶片组织,然后将取下的长条形叶片组织在 2.5% 戊二醛（用 pH=7.2 的磷酸缓冲液配制）中于 22℃下至少固定 2 h 以上,再在相同温度下用锇酸（OsO_4）固定 2 h,之后分别在 30%、50%、70% 和 90% 丙酮中进行脱水处理,然后浸泡于环氧树脂 812 中 2 h,并使其聚合 8 h,最后制备超薄切片,并利用透射电镜 H600-Ⅳ 观察拍片。

§4.2 生理指标测定

4.2.1 叶水势的测定

2011年7月7日和8日黎明，选择健康完整的第三或第四叶位叶片（每个处理选择3张叶片），用锋利的剪刀将叶片剪下后立即密封入装有湿润滤纸的塑料袋，并置于冰盒中迅速带回实验室，使用WP4露点水势仪（Decagon Devices，Inc.，Pullman，WA，USA）进行黎明前叶水势测定。

4.2.2 碳同位素分馏的测定

测定碳同位素分馏（Δ）的叶片与气体交换测量的叶片为同一张，具体方法参照Li等（2000）。将叶片在80℃下烘干约48 h直至恒重，在研钵中研磨成细粉，过100目的筛子，将过筛后的细粉收集起来用于碳同位素组成的测定。利用质谱仪测定燃烧样品的稳定碳同位素丰度（Finnegan MAT Delta－E，Bremen，Germany）。δ的总精度大于0.1‰。测定分析在中国科学院生态和环境科学稳定同位素实验室（SILEER）完成。然后参照Farquhar等（1989）的公式计算碳同位素分馏：$\Delta = (\delta_a - \delta_p)/(1 + \delta_p)$，$\delta_a$和$\delta_p$分别是空气和植物材料的碳同位素组成。本书试验研究$\delta_a$假定为$-8.0$‰。稳定碳同位素组成值的计算参考国际标准VPDB。

4.2.3 叶、茎、根的铅含量测定

用去离子水彻底冲洗叶、茎和根，然后在80℃下干燥直至恒重，将烘干的样品研磨成粉末，过100目的筛子，筛好的样品

粉末置于硝酸-高氯酸（3∶1，V/V）中进行消化酸解，最后用石墨炉原子吸收光谱法测定铅含量。

4.2.4 修复相关参数的计算

4.2.4.1 转移系数（T_f）

转移系数用于计算植物将重金属从根部转移到地上部分的能力。转移系数高，有利于植物修复，计算公式如下：

$$T_f = （叶片铅含量/根铅含量）\times 100\%$$

4.2.4.2 容忍系数（T_i）

容忍系数的计算公式为

$$T_i = 重金属处理植物的生物量/对照植物的生物量$$

$T_i > 1$，说明生物量呈现净增长，表明植物有较好的重金属容忍性；$T_i < 1$，说明生物量呈现净减少，表明植物生长受到胁迫；$T_i = 1$，表明重金属胁迫植物与对照相比没有明显变化。

4.2.4.3 生物浓度因子（BCF）

生物浓度因子定义为植物当中的金属浓度与土壤当中金属浓度的比值。该参数提供了植物对金属的吸收信息，用于估算植物的重金属吸收效率。$BCF > 1$，表明植物为潜在的重金属超积累种。其计算公式为

$$BCF = 植物组织铅含量/土壤铅含量$$

第5章 试验数据统计分析

§5.1 干旱和铅试验数据统计分析

使用 SPSS16.0 统计软件（SPSS Inc., Chicago, IL, USA）对所测数据进行多因素方差分析，以确定性、干旱和铅（Pb_{soil} 或 Pb_{leaf}）及其交互作用的影响，并对以上各形态、生理和生化指标进行一元方差分析（ANOVA），平均数间的多重比较采用 Tukey 检验（$P<0.05$）。在方差分析之前，首先检测数据是否成正态分布、方差是否齐性，否则需要对数据进行对数转换。

§5.2 干旱和嫁接试验数据统计分析

将所有数据分成 2 组进行统计分析，组 1 包括 100% 田间持水量处理的所有个体，组 2 包括 30% 田间持水量处理的所有个体。使用 SPSS16.0 统计软件（SPSS Inc., Chicago, IL, USA）对每组数据分别进行一元方差分析，平均数间的多重比较采用 Tukey 检验（$P<0.05$）。在方差分析之前，检测数据是否成正态分布、方差是否齐性，否则需要对数据进行对数转换。

另外，采用独立样本 t 检验评估每个嫁接组合的干旱组和对照组间的差异显著性。最后，采用二元方差分析评估水分处理和嫁接组合的交互影响。

研究结果与分析篇

第 6 章 青杨雌雄植株对干旱胁迫、铅胁迫及交互胁迫的响应差异

§6.1 主要研究结果

6.1.1 干旱胁迫、铅胁迫及交互胁迫下青杨雌雄植株的干物质积累和分配差异

干旱胁迫、铅胁迫及交互胁迫下青杨雌雄植株生物量的积累和分配如图 6-1 所示。Pb_{soil} 单独处理并没有导致雌雄植株生物量累积的明显下降 [图 6-1(a)]。与对照组相比,两性植株的根茎比(R/S)均明显下降 [图 6-1(c)]。Pb_{soil} 和干旱的组合处理明显减少了两性植株的叶生物量、茎生物量、根生物量以及总生物量,相比于雄性植株,雌性植株减少得更加明显 [图 6-1(a)]。另外,Pb_{leaf} 单独处理及 Pb_{leaf} 和干旱组合处理明显减少了两性的叶生物量、茎生物量、根生物量以及总生物量(Pb_{leaf} 单独处理下的雄性茎生物量除外),相比于雄性植株,雌性植株减少得更加明显 [图 6-1(b)]。两性的 R/S 几乎未受到 Pb_{leaf} 单独处理的影响。然而,Pb_{leaf} 和干旱组合处理明显增加了两性植株的 R/S [图 6-1(d)]。在 Pb_{leaf} 单独处理及 Pb_{leaf} 和干旱组合处理中,根生物量和总生物量存在明显的性别差异,相

比于雌性植株，雄性植株积累了更高的根生物量和总生物量[图 6-1(b)]。

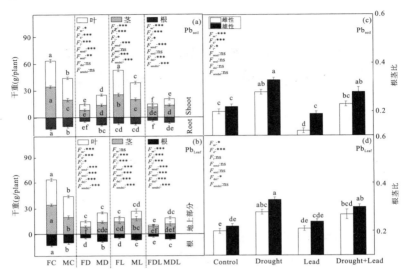

图 6-1　在干旱胁迫、铅胁迫及交互胁迫下青杨雌雄植株生物量的积累和分配

注：1. (a)(b) 植物不同部位的干物质积累；(c)(d) 根茎比。

2. 每个柱上不同的小写字母表示处理间在 $P<0.05$ 水平上差异显著（Tukey 检验）。数值为平均值±标准误差，$n=5$。*，$0.01<P<0.05$；**，$0.001<P<0.01$；***，$P\leqslant 0.001$；ns，差异不显著。

3. F_{se}，性别影响；F_d，干旱影响；F_l，铅影响；$F_{se\times d}$，性别和干旱交互影响；$F_{se\times l}$，性别和铅交互影响；$F_{d\times l}$，干旱和铅交互影响；$F_{se\times d\times l}$，性别、干旱和铅交互影响；FC，雌性对照；MC，雄性对照；FD，干旱处理的雌性植株；MD，干旱处理的雄性植株；FL，铅处理的雌性植株；ML，铅处理的雄性植株；FDL，铅和干旱处理的雌性植株；MDL，铅和干旱处理的雄性植株。

6.1.2 干旱胁迫、铅胁迫及交互胁迫下青杨雌雄植株的气体交换和用水效率差异

与对照组相比，Pb_{soil} 单独处理下，两性的 Δ 均增加。相比于雄性植株，雌性植株 Δ 增加更加明显 [图 6-2(g)]。反之，在 Pb_{soil} 单独处理下，两性植株的 P_n（雄性植株的 P_n 除外）、T_{Chl} 及 Chla/Chlb 几乎未受影响 [图 6-2(a)、图 6-2(e)、图 6-2(c)]。在 Pb_{soil} 和干旱组合处理下，雌性植株的 P_n、Δ、T_{Chl} 及 Chla/Chlb 分别下降了 79.7%、3.8%、68.8% 及 15.5%。对比之下，在 Pb_{soil} 和干旱组合处理下，雄性植株的 P_n、Δ、T_{Chl} 及 Chla/Chlb 只下降了 42.5%、4.0%、49.2% 及 4.0% [图 6-2(a)、图 6-2(g)、图 6-2(e)、图 6-2(c)]。另外，与对照组相比，Pb_{leaf} 单独处理明显减少了两性植株的 P_n [图 6-2(b)] 和 T_{Chl} [图6-2(f)]，增加了两性植株的 Δ [图 6-2(h)]，而两性植株的 Chla/Chlb 几乎未受影响 [图 6-2(d)]。Pb_{leaf} 和干旱组合处理导致雌性植株的 P_n、Δ、T_{Chl} 及 Chla/Chlb 分别下降了 87.8%、0.9%、74.3% 及 28.8%。对比之下，在 Pb_{leaf} 和干旱组合处理下，雄性植株的 P_n、Δ、T_{Chl} 及 Chla/Chlb 只下降了 55.9%、3.4%、54.2% 和 11.8% [图 6-2(b)、图 6-2(h)、图 6-2(f)、图6-2(d)]。

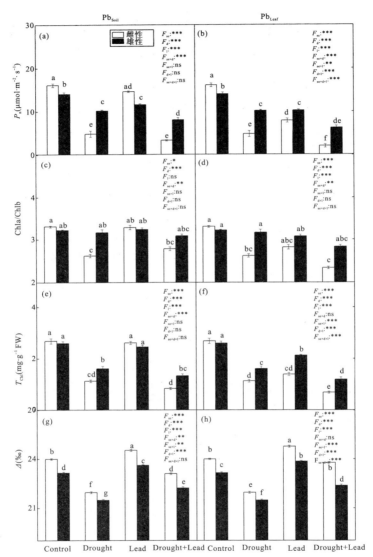

图 6-2 干旱胁迫、铅胁迫及交互胁迫下青杨雌雄植株的净光合作用速率（P_n）、总叶绿素含量、叶绿素 a/b 及碳同位素分馏

注：1. 每个柱上不同的小写字母表示处理间在 $P<0.05$ 水平上差异

第6章 青杨雌雄植株对干旱胁迫、铅胁迫及交互胁迫的响应差异

显著（Tukey 检验）。数值为平均值±标准误差，$n=5$。$*$，$0.01<P<0.05$；$**$，$0.001<P<0.01$；$***$，$P\leqslant 0.001$；ns，差异不显著。

2. F_{se}，性别影响；F_d，干旱影响；F_l，铅影响；$F_{se\times d}$，性别和干旱交互影响；$F_{se\times l}$，性别和铅交互影响；$F_{d\times l}$，干旱和铅交互影响；$F_{se\times d\times l}$，性别、干旱和铅交互影响。

光响应曲线和 CO_2 响应曲线分析所得参数如表 6-1、表 6-2、图 6-3 及图 6-4 所示。在所有的处理中，两性植株的 A_{max}、L_{sp} 和 CE 均明显下降，在 Pb_{leaf} 单独处理及 Pb_{leaf} 和干旱组合处理下，雌性植株的 A_{max}、L_{sp} 和 CE 的下降尤其明显（表 6-1 和表 6-2）。对比之下，在所有的处理中，L_{CP}、R_d 和 $\varGamma$ 则明显增加。相比于雄性植株，雌性植株的 L_{CP}、R_d 和 $\varGamma$ 增加得更加明显，尤其是在 Pb_{leaf} 单独处理及 Pb_{leaf} 和干旱组合处理下。但是在所有处理中，两性植株的 AQE 均几乎未受影响。

表 6-1 干旱胁迫、铅胁迫及交互胁迫下青杨雌雄植株的表观量子产量（AQE）、最大净光合作用速率（A_{max}）、高光饱和点（L_{sp}）、低光补偿点（L_{cp}）及暗呼吸速率（R_d）

Pb$_{soil}$/(mg·kg^{-1} dry soil)	田间持水量/%	性别	AQE	A_{max}	L_{sp}	L_{cp}	R_d
0	100	雌性	0.0422±0.005ab	48.48±3.16b	1163.33±46.32e	14.52±0.14f	0.552±0.06g
0	100	雄性	0.0410±0.002ab	51.57±2.23a	1264.13±61.13d	6.32±0.09g	0.275±0.04h
0	50	雌性	0.0219±0.003ab	18.26±1.87e	872.87±38.74g	39.08±0.16c	1.468±0.02b
0	50	雄性	0.0331±0.005ab	47.56±3.59b	1471.24±66.93b	34.38±0.12d	0.959±0.04f
900	100	雌性	0.0458±0.002a	39.98±2.21c	914.01±33.35f	41.02±0.19c	1.325±0.05c
900	100	雄性	0.0390±0.001ab	52.34±3.75a	1370.57±49.09c	28.65±0.11e	1.079±0.04e
900	50	雌性	0.0223±0.002ab	14.31±1.94f	707.54±32.17h	65.83±1.01b	2.291±0.06a
900	50	雄性	0.0176±0.004ab	28.23±2.25d	1677.07±68.83a	73.09±1.15a	1.124±0.03d
P		F_{se}	ns	***	***	***	***
		F_d	***	***	***	***	***
		F_1	ns	***	***	***	***

第6章 青杨雌雄植株对干旱胁迫、铅胁迫及交互胁迫的响应差异

续表

Pb	田间持水量/%	性别	AQE	A_{max}	L_{sp}	L_{cp}	R_d
P		$F_{se \times d}$	ns	***	***	***	***
		$F_{se \times l}$	ns	***	***	***	***
		$F_{d \times l}$	ns	***	***	***	***
		$F_{se \times d \times l}$	ns	***	***	***	***
Pb$_{soil}$/(mg·kg^{-1} dry soil)							
0	100	雌性	0.0422±0.005ab	48.48±3.16b	1163.33±46.32e	14.52±0.14f	0.552±0.06g
0	100	雄性	0.0410±0.002ab	51.57±2.23a	1264.13±61.13d	6.32±0.09g	0.275±0.04h
0	50	雌性	0.0219±0.003ab	18.26±1.87e	872.87±38.74g	39.08±0.16c	1.468±0.02c
0	50	雄性	0.0331±0.005ab	47.56±3.59b	1471.24±66.93b	34.38±0.12d	0.959±0.04d
600	100	雌性	0.0279±0.002a	31.14±1.76d	1160.16±34.11f	44.03±1.07b	1.527±0.04b
600	100	雄性	0.0307±0.003a	37.63±2.28c	1247.47±71.61d	21.74±0.45e	0.770±0.01f
600	50	雌性	0.0187±0.001a	10.69±1.43f	666.60±11.39h	94.95±1.16a	2.737±0.05a
900	50	雄性	0.0210±0.004a	36.89±2.25c	1794.57±77.52a	37.90±0.53c	0.914±0.02e

Pb$_{leaf}$/(mg/plant)

续表

Pb	田间持水量/%	性别	AQE	A_{max}	L_{sp}	L_{cp}	R_d
$Pb_{leaf}/(mg/plant)$							
P		F_{se}	ns	***	***	***	***
		F_d	**	***	***	***	***
		F_l	ns	***	***	***	***
		$F_{se×d}$	ns	***	***	***	***
		$F_{se×l}$	ns	ns	***	***	***
		$F_{d×l}$	ns	***	***	***	***
		$F_{se×d×l}$	ns	***	***	***	***

注：1. 每个数值均是平均值±标准误差 ($n=5$)。每列不同的小写字母表示处理间在 $P<0.05$ 水平上差异显著（Tukey检验）。*, $0.01<P<0.05$; **, $0.001<P<0.01$; ***, $P\leq0.001$; ns, 差异不显著。

2. F_{se}, 性别影响; F_d, 干旱影响; F_l, 铅影响; $F_{se×d}$, 性别和干旱交互影响; $F_{se×l}$, 性别和铅交互影响; $F_{d×l}$, 干旱和铅交互影响; $F_{se×d×l}$, 性别、干旱和铅交互影响。

表6-2 干旱胁迫、铅胁迫及交互胁迫下青杨雌雄植株的羧化效率（CE）和CO_2补偿点（Γ）

Pb	田间持水量/%	性别	CE	Γ
Pb_{soil}/(mg·kg^{-1} dry soil)				
0	100	雌性	0.0947±0.005a	50.88±2.09d
0	100	雄性	0.0909±0.002a	59.33±2.38c
0	50	雌性	0.0415±0.003cd	69.72±3.11a
0	50	雄性	0.0641±0.006bc	64.27±2.24b
900	100	雌性	0.0757±0.003ab	58.55±3.16c
900	100	雄性	0.0633±0.005bc	62.43±4.38b
900	50	雌性	0.0349±0.002d	69.44±3.74a
900	50	雄性	0.0525±0.004bcd	64.56±2.19b
P		F_{se}	ns	ns
		F_d	***	***
		F_l	**	***
		$F_{se×d}$	**	***
		$F_{se×l}$	ns	*
		$F_{d×l}$	ns	***
		$F_{se×d×l}$	ns	***
Pb_{leaf}/(mg/plant)				
0	100	雌性	0.0947±0.005a	50.88±2.09e
0	100	雄性	0.0909±0.002a	59.33±2.38d
0	50	雌性	0.0415±0.003cd	69.72±3.11b
0	50	雄性	0.0641±0.006bc	64.27±2.24c
600	100	雌性	0.0417±0.006bcd	60.55±2.69d

续表

Pb	田间持水量/%	性别	CE	Γ
Pb$_{leaf}$/(mg/plant)				
600	100	雄性	0.0654±0.003ab	59.39±3.50d
600	50	雌性	0.0121±0.001d	77.17±2.22a
600	50	雄性	0.0393±0.002cd	68.53±4.16b
P	F_{se}		***	***
	F_d		***	***
	F_l		***	***
	$F_{se×d}$		ns	***
	$F_{se×l}$		ns	***
	$F_{d×l}$		ns	ns
	$F_{se×d×l}$		ns	**

注：1. 每个数值均是平均值±标准误差（$n=5$）。每列不同的小写字母表示处理间在 $P<0.05$ 水平上差异显著（Tukey 检验）。*，$0.01<P<0.05$；**，$0.001<P<0.01$；***，$P\leqslant 0.001$；ns，差异不显著。

2. F_{se}，性别影响；F_d，干旱影响；F_l，铅影响；$F_{se×d}$，性别和干旱交互影响；$F_{se×l}$，性别和铅交互影响；$F_{d×l}$，干旱和铅交互影响；$F_{se×d×l}$，性别、干旱和铅交互影响。

第6章 青杨雌雄植株对干旱胁迫、铅胁迫及交互胁迫的响应差异

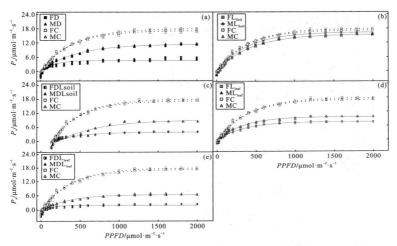

图6-3 干旱胁迫、铅胁迫及交互胁迫下青杨雌雄植株的光响应曲线

注:1. 每个符号代表一个测量值。回归线拟合到数据上($n=5$)。

2. FC,雌性植株对照;MC,雄性植株对照;FD,干旱处理的雌性植株;MD,干旱处理的雄性植株;FL_{soil},Pb_{soil}处理的雌性植株;ML_{soil},Pb_{soil}处理的雄性植株;FDL_{soil},Pb_{soil}和干旱处理的雌性植株;MDL_{soil},Pb_{soil}和干旱处理的雄性植株;FL_{leaf},Pb_{leaf}处理的雌性植株;ML_{leaf},Pb_{leaf}处理的雌性植株;FDL_{leaf},Pb_{leaf}和干旱处理的雌性植株;MDL_{leaf},Pb_{leaf}和干旱处理的雄性植株。

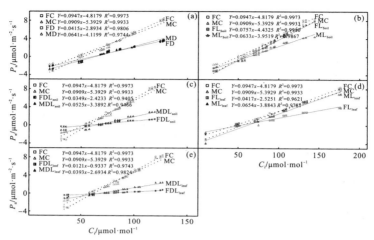

图 6-4 干旱胁迫、铅胁迫及交互胁迫下青杨雌雄植株的 CO_2 响应曲线（0～200 $\mu mol \cdot mol^{-1}$ CO_2 浓度范围内，P_n 和 C_i 的线性回归）

注：1. 每个符号代表一个测量值。回归线拟合到数据上（$n = 5$）。

2. FC，雌性植株对照；MC，雄性植株对照；FD，干旱处理的雌性植株；MD，干旱处理的雄性植株；FL_{soil}，Pb_{soil} 处理的雌性植株；ML_{soil}，Pb_{soil} 处理的雄性植株；FDL_{soil}，Pb_{soil} 和干旱处理的雌性植株；MDL_{soil}，Pb_{soil} 和干旱处理的雄性植株；FL_{leaf}，Pb_{leaf} 处理的雌性植株；ML_{leaf}，Pb_{leaf} 处理的雄性植株；FDL_{leaf}，Pb_{leaf} 和干旱处理的雌性植株；MDL_{leaf}，Pb_{leaf} 和干旱处理的雄性植株。

6.1.3 干旱胁迫、铅胁迫及交互胁迫下青杨雌雄植株的叶绿素荧光和淬灭参数差异

在所有的处理中，两性植株的 PSⅡ 光化学量子产量（$\Phi_{PSⅡ}$）、光化学淬灭（q_P）和 PSⅡ 开放反应中心的光量子效率（F_v/F_m）明显下降，尤其是 Pb_{leaf} 处理的雌性植株（图 6-5）。对比之下，在所有的处理中，非光化学淬灭（q_N）明显增加，Pb_{leaf} 处理的雌性植株增加尤其明显［图 6-5(e)、图 6-5(f)］。

第6章 青杨雌雄植株对干旱胁迫、铅胁迫及交互胁迫的响应差异

在 Pb_{soil} 单独处理中，雌性植株的 Φ_{PSII}、q_P 和 F_v/F_m 下降了 7.2%、3.0% 及 1.5%，反之雄性植株的 Φ_{PSII}、q_P 和 F_v/F_m 只下降了 3.4%、2.0% 和 0.7% [图 6-5(g)、图 6-5(c)、图 6-5(a)]。另外，在 Pb_{leaf} 单独处理中，雌性植株的 Φ_{PSII}、q_P 和 F_v/F_m 分别下降了 24.4%、11.1% 和 3.3%，反之雄性的 Φ_{PSII}、q_P 和 F_v/F_m 分别下降了 8.6%、3.6% 和 1.6% [图 6-5(h)、图 6-5(d)、图 6-5(b)]。

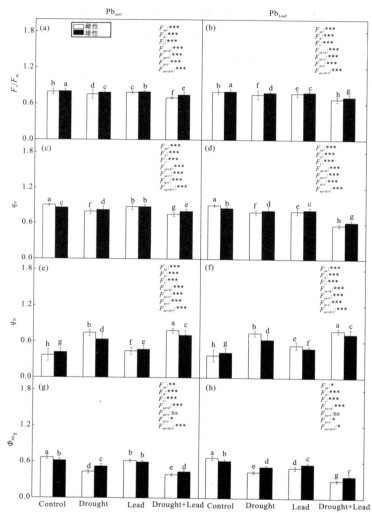

图 6-5 干旱胁迫、铅胁迫及交互胁迫下青杨雌雄植株的 PSⅡ开放反应中心的光量子效率（F_v/F_m）、光化学淬灭（q_P）、非光化学淬灭（q_N）及 PSⅡ光化学量子产量（$\Phi_{PSⅡ}$）

注：1. 每个柱上不同的小写字母表示处理间在 $P<0.05$ 水平上差异显著（Tukey 检验）。数值为平均值±标准误差，$n=5$。*，$0.01<P<0.05$；

，$0.001 < P < 0.01$；*，$P \leqslant 0.001$；ns，差异不显著。

2. F_{se}，性别影响；F_d，干旱影响；F_l，铅影响；$F_{se \times d}$，性别和干旱交互影响；$F_{se \times l}$，性别和铅交互影响；$F_{d \times l}$，干旱和铅交互影响；$F_{se \times d \times l}$，性别、干旱和铅交互影响。

6.1.4 干旱胁迫、铅胁迫及交互胁迫下青杨雌雄植株的铅吸收和转移差异

在 Pb_{soil} 单独处理及 Pb_{soil} 和干旱组合处理中，雌雄植株根部均吸收了大量的铅，而在这两种处理下，雌性植株茎中的铅含量大约分别是雄性植株茎中铅含量的 18.4 倍和 11.5 倍。在 Pb_{soil} 单独处理中，雄性植株和雌性植株的叶片铅含量与根部铅含量均未表现出明显的性别间差异。然而，在 Pb_{soil} 和干旱组合处理中，雄性植株和雌性植株叶片铅含量与根部铅含量存在明显的性别间差异。另外，在 Pb_{leaf} 单独处理及 Pb_{leaf} 和干旱组合处理中，雄性植株和雌性植株叶与茎当中的铅含量存在明显的性别间差异。在 Pb_{leaf} 单独处理及 Pb_{leaf} 和干旱组合处理中，雌性植株叶片铅含量分别是雄性植株叶片铅含量的 2.3 倍和 2.1 倍。在 Pb_{leaf} 单独处理及 Pb_{leaf} 和干旱组合处理中，与叶片相比，雄性植株和雌性植株的茎与根只吸收了很少量的铅（表 6-3）。

表6-3 干旱胁迫、铅胁迫及交互胁迫下青杨雌雄植株叶、茎和根的铅含量

Pb_{soil}/(mg·kg^{-1} dry soil)	田间持水量/%	性别	叶的铅含量/(mg·kg^{-1})	茎的铅含量/(mg·kg^{-1})	根的铅含量/(mg·kg^{-1})
0	100	雌性	1.36±0.05cd	1.36±0.08c	4.20±0.09c
0	100	雄性	1.79±0.09c	1.04±0.10cd	4.21±0.14c
0	50	雌性	2.49±0.15b	0.81±0.05d	1.91±0.04e
0	50	雄性	1.19±0.06d	0.55±0.04e	3.43±0.23d
900	100	雌性	4.68±0.09a	126.93±13.36a	754.25±23.84a
900	100	雄性	5.02±0.13a	6.89±0.46b	753.67±30.76a
900	50	雌性	5.12±0.10a	96.21±1.92a	716.04±5.28a
900	50	雄性	2.87±0.12b	8.38±0.75b	509.03±17.27b
P		F_{se}	***	***	*
		F_d	***	***	***
		F_l	***	***	***
		$F_{se×d}$	***	ns	*

续表

Pb	田间持水量/%	性别	叶的铅含量/(mg·kg^{-1})	茎的铅含量/(mg·kg^{-1})	根的铅含量/(mg·kg^{-1})
Pb$_{soil}$/(mg·kg^{-1} dry soil)					
		$F_{se×1}$	**	***	***
P		$F_{d×1}$	***	***	***
		$F_{se×d×1}$	*	*	***
0	100	雌性	1.36±0.05ef	1.36±0.08c	4.20±0.09ab
0	100	雄性	1.79±0.09de	1.04±0.10c	4.21±0.14a
0	50	雌性	2.49±0.15d	0.81±0.05e	1.91±0.04d
0	50	雄性	1.19±0.06f	0.55±0.04d	3.43±0.23c
600	100	雌性	67.71±5.47a	3.27±0.13a	5.95±0.10a
600	100	雄性	28.92±1.63b	2.07±0.01ab	5.28±0.15a
600	50	雌性	22.54±2.42b	1.96±0.02d	3.15±0.05d
600	50	雄性	10.56±0.86c	1.08±0.01bc	4.48±0.23bc
Pb$_{leaf}$/(mg/plant)					

续表

Pb	田间持水量/%	性别	叶的铅含量/(mg·kg^{-1})	茎的铅含量/(mg·kg^{-1})	根的铅含量/(mg·kg^{-1})
Pb$_{leaf}$/(mg/plant)					
	P	F_{se}	***	***	***
		F_d	***	***	***
		F_l	***	***	*
		$F_{se×d}$	***	ns	***
		$F_{se×l}$	***	**	*
		$F_{d×l}$	***	ns	ns
		$F_{se×d×l}$	***	ns	ns

注：1. 每个数值均是平均值±标准误差（$n=5$）。每列不同的小写字母表示处理间在 $P<0.05$ 水平上差异显著（Tukey 检验）。*, $0.01<P<0.05$；**, $0.001<P<0.01$；***, $P\leqslant0.001$；ns，差异不显著。

2. F_{se}，性别影响；F_d，干旱影响；F_l，铅影响；$F_{se×d}$，性别和干旱交互影响；$F_{se×l}$，性别和铅交互影响；$F_{d×l}$，干旱和铅交互影响；$F_{se×d×l}$，性别、干旱和铅交互影响。

6.1.5 干旱胁迫、铅胁迫及交互胁迫下青杨雌雄植株的重金属修复能力差异

雌雄植株根部和地上部分的生物浓度因子（BCF）见表6-4。在Pb_{soil}单独处理及Pb_{soil}和干旱组合处理中，雌性植株地上部分的BCF分别占整个植株BCF的15%和12%，对比之下，雄性植株仅分别占1%和2%。这表明在Pb_{soil}单独处理及Pb_{soil}和干旱组合处理中，相比于雄性植株，雌性植株的叶和茎积累铅的能力更高。通过转移系数（T_f）的计算，可以进一步评估雌雄植株在地上部分积累铅的能力。转移系数表示地上部分铅含量与根部铅含量的百分比，在Pb_{soil}单独处理中，雌雄植株间的转移系数没有明显的差异。但是在Pb_{soil}和干旱组合处理中，雌性植株T_f明显高于雄性植株T_f（表6-4）。容忍系数（T_i）是另一个评估植物修复能力的重要参数。在Pb_{soil}单独处理中，雄性植株和雌性植株的T_i没有明显的性别间差异。反之，在Pb_{soil}和干旱组合处理中，雄性植株的T_i明显比雌性植株高。

表6-4 干旱胁迫、铅胁迫及交互胁迫下青杨雌雄植株的转移系数（T_f）、容忍系数（T_i）和生物浓度因子（BCF）

Pb_{soil}/ (mg·kg^{-1} dry soil)	田间持水量/%	性别	T_f	T_i	BCF_{shoot}	BCF_{root}
900	100	雌性	0.62±0.02ab	0.78±0.04a	0.15±0.01a	0.84±0.03a
900	100	雄性	0.67±0.05ab	0.85±0.05a	0.01±0.01b	0.84±0.03a
900	50	雌性	0.72±0.02a	0.24±0.03c	0.11±0.01a	0.80±0.01a
900	50	雄性	0.56±0.01b	0.49±0.02b	0.01±0.01b	0.57±0.02b
P		F_{se}	ns	**	***	***
		F_d	ns	***	*	***
		$F_{se×d}$	**	*	ns	

注：1. 每个数值均是平均值±标准误差（n=5）。每列不同的小写字母表示处理间在 P<0.05 水平上差异显著（Tukey 检验）。*，0.01<P<0.05；**，0.001<P<0.01；***，P≤0.001；ns，差异不显著。

2. F_{se}，性别影响；F_d，干旱影响；F_l，铅影响；$F_{se×d}$，性别和干旱交互影响；$F_{se×l}$，性别和铅交互影响；$F_{d×l}$，干旱和铅交互影响；$F_{se×d×l}$，性别、干旱和铅交互影响。

6.1.6 干旱胁迫、铅胁迫及交互胁迫下青杨雌雄植株的细胞超微结构差异

雄性植株和雌性植株在细胞超微结构上的差异如图6-6所示。在对照组中，两性植株的叶绿体均呈典型的透镜形状，内膜系统排列规则。基粒和基质内囊体沿着主轴线性排列，线粒体呈现典型的结构和清晰的嵴 [图6-6(a)、图6-6(b)]。雌雄植株的叶片细胞均有光滑、清晰和持续的细胞膜及细胞壁，此外，颗粒状的细胞质密集填充于各细胞器，核结构典型，核仁和核膜清晰可见，染色质分布均匀 [图6-6(a)、图6-6(b)]。Pb_{soil} 单独处理几乎没有对雄性植株和雌性植株的叶绿体和线粒体造成影响 [图6-6(e)、图6-6(f)]。但是在 Pb_{soil} 和干旱组合处理中，两性植株叶细胞的超微结构均受到严重的破坏 [图6-6(g)、图6-6(h)]，具体表现为细胞壁非正常加厚，叶绿体结构紊乱并分布有较大的质体小球及线粒体嵴降解模糊 [图6-6(g)、图6-6(h)]。在 Pb_{soil} 和干旱组合处理中，与雄性植株相比，雌性植株叶肉细胞中正常叶绿体的数量更少且多数扭曲变形。与 Pb_{soil} 处理相比，Pb_{leaf} 处理对两性植株叶细胞的超微结构有更加严重的影响 [图6-6(i)、图6-6(j)]，两性植株尤其是雌性植株的细胞质空泡化明显且细胞器密度明显下降。相比于 Pb_{soil} 和干旱组合处理 [图6-6(g)、图6-6(h)]，Pb_{leaf} 和干旱组合处理 [图6-6(k)、图6-6(l)] 对细胞器的有害影响更加明显。在 Pb_{leaf} 和干旱组合处理中，相比于雄性植株，雌性植株叶细胞超微结构的损坏更加明显。在 Pb_{leaf} 和干旱组合处理中，雌性植株的叶绿体外形相当畸形，内膜系统排列极不规则，基粒非常大且排列不规则，并随机分布于基质中。此外，染色质浓缩，核仁不清晰，核膜降解明显 [图6-6(l)]。

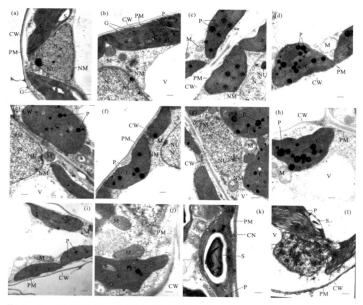

图6-6 干旱胁迫、铅胁迫及交互胁迫下青杨雌雄植株叶肉细胞的透射电子显微镜观察

注：(a) 雄性对照的叶肉细胞；(b) 雌性对照的叶肉细胞；(c) 干旱处理的雄性植株的叶肉细胞；(d) 干旱处理的雌性植株的叶肉细胞；(e) Pb_{soil}处理的雄性植株的叶肉细胞；(f) Pb_{soil}处理的雌性植株的叶肉细胞；(g) Pb_{soil}和干旱组合处理的雄性植株的叶肉细胞；(h) Pb_{soil}和干旱组合处理的雌性植株的叶肉细胞；(i) Pb_{leaf}处理的雄性植株的叶肉细胞；(j) Pb_{leaf}处理的雌性植株的叶肉细胞；(k) Pb_{leaf}和干旱组合处理的雄性植株的叶肉细胞；(l) Pb_{leaf}和干旱组合处理的雌性植株的叶肉细胞。

§6.2 讨论

在中国，随着大规模矿业和工业活动的开展，重金属的污染越来越严重，如何处理农业、林业和生态系统的污水、污泥及土壤酸化已经是一个日益严重的问题。重金属毒性可以对植物产生很多直接和间接的影响。在本研究中，我们评估了干旱和铅对雌雄异株青杨的一些重要形态特征和生理生化特征的影响，具体包括干物质积累和分配、气体交换参数、叶绿素荧光、铅吸收和转移以及细胞的超微结构。总的来看，Pb_{soil}单独处理对雄性植株和雌性植株的影响很小。相反，Pb_{leaf}单独处理、Pb_{soil}和干旱组合处理以及Pb_{leaf}和干旱组合处理对两性植株的影响较大，尤其是对雌性植株的影响更大。

6.2.1 青杨雌雄植株的干物质积累和分配的变化

植物对重金属胁迫的一个最初响应就是生物量积累下降，它可以作为评估植物重金属容忍性的一项指标。Pb_{soil}单独处理导致青杨雄雌植株的干物质积累略微下降，与对照组相比，雄雌植株分别下降了 16.6%、22.1% [图 6-1(a)]；Pb_{leaf}单独处理明显减少了雄雌植株的干物质积累，与对照组相比，雄雌植株分别下降了 49.1% 和 68.9% [图 6-1(b)]。Pb_{soil} 和 Pb_{leaf}处理均导致雌性植株和雄性植株的根生物量显著下降。但是在 Pb_{leaf}单独处理下，根生物量的减少并没有导致根茎比（R/S）的减少[图 6-1(d)]；在 Pb_{soil}单独处理下，R/S 则明显减少 [图 6-1(c)]。Oncel 等（2000）的研究也表明，在铅或其他重金属添加到土壤的条件下，植物的 R/S 明显下降。

6.2.2 青杨雌雄植株的气体交换和水分利用效率的变化

重金属通过损伤光合作用相关细胞器（如叶绿体、气孔器）及水传导系统而影响 CO_2 的吸收和固定。在本研究所测量的光合作用相关参数中，净光合作用速率受铅的影响最大，尤其是在 Pb_{leaf}单独处理、Pb_{soil}和干旱组合处理及 Pb_{leaf}和干旱组合处理下

[图6-2(a)、图6-2(b)]。在 Pb_{leaf} 单独处理下，雄性植株和雌性植株的 P_n 明显下降，与对照组相比分别下降了 27.42% 和 51.12% [图6-2(b)]。对比之下，在 Pb_{soil} 单独处理下，雄性植株和雌性植株的 P_n 只略微下降，与对照组相比分别下降了 17.31% 和 9.24% [图6-2(a)]。这可能是因为相比于 Pb_{soil} 处理，在 Pb_{leaf} 处理下，雌雄植株叶片中累积了大量的铅（表6-3）。叶片中累积的 Pb 也可能影响了叶绿素的合成。在 Pb_{leaf} 单独处理以及 Pb_{leaf} 和干旱组合处理下，雄性植株和雌性植株的叶绿素均减少，但相比于叶片铅含量较高的雌性植株，叶片铅含量较低的雄性植株中叶片组织的叶绿素含量更多 [图6-2(f)、图6-2(e)]。因此我们推断，在 Pb_{leaf} 单独处理以及 Pb_{leaf} 和干旱组合处理下，相比于雌性植株，雄性植株叶绿素的生物合成受到的影响更小，进而有助于其在逆境条件下维持较高的光合作用速率。

重金属通常会扰乱植物的水分平衡，这被认为是重金属毒性的一个最初的且很重要的原因。我们的研究也发现，重金属干扰了雌雄植株的水分平衡。Pb_{soil} 单独处理和 Pb_{leaf} 单独处理的雄雌植株的长期用水效率均下降 [图6-2(g)、图6-2(h)]，这与先前的研究一致。先前的研究发现，生长在铜污染土壤中的小麦比生长在没有铜污染土壤中的小麦的用水效率低。也有研究发现，不同的重金属对同一种植物的蒸腾速率和生长有不同的影响。在相近水平的叶片重金属浓度下，铅导致用水效率急剧下降，而铬对 CO_2 吸收和蒸腾的抑制程度几乎相同，并没有改变用水效率。我们的研究表明，铅对用水效率也有明显的影响，有趣的是，不同的铅处理方式对用水效率有不同的影响。Pb_{leaf} 处理的雄性和雌性的用水效率明显比 Pb_{soil} 处理的低 [图6-2(h)、图6-2(g)]。

6.2.3　青杨雌雄植株的叶绿素荧光和淬灭参数的变化

叶绿素荧光能够监测 PSⅡ 的功能和光捕获效率。对铅胁迫

第6章　青杨雌雄植株对干旱胁迫、铅胁迫及交互胁迫的响应差异

植物的叶绿素荧光参数的分析可以进一步获得铅诱导损伤机制。叶绿素荧光和淬灭参数的测定能够检测出铅诱导损伤的一系列改变，包括光合作用电子传递链氧化还原状态（光化学荧光淬灭，q_P）的改变和类囊体能量化（非光化学荧光淬灭，q_N）的改变。在本研究的所有处理中，光化学淬灭均下降，表明逆境胁迫下PSⅡ反应中心关闭比例增加［图6-3(d)、图6-3(c)］。另外，如图6-3所示，在Pb_{soil}单独处理条件下，两性植株的光化学淬灭参数并没有明显的差异，但是在Pb_{leaf}单独处理及Pb_{leaf}和干旱组合处理下，雄性植株比雌性植株具有更高的光化学淬灭［图6-3(d)、图6-3(c)］，这表明在Pb_{leaf}单独处理及Pb_{leaf}和干旱组合处理下，相比于雄性植株，雌性植株的PSⅡ反应中心关闭得更多。在所有的处理中，非光化学淬灭均增加［图6-3(e)、图6-3(f)］，反映了类囊体能量化程度，例如在组合胁迫下，跨类囊体膜的质子梯度增加尤为明显。总的来看，q_P和q_N分析显示，所有处理均影响了PSⅡ原初电子受体的氧化还原状态和开放的PSⅡ反应中心的激发能捕获效率。与Pb_{soil}单独处理相比，在Pb_{leaf}单独处理下，光化学淬灭、激发能捕获效率和PSⅡ电子运输效率下降更明显，并且雌性植株比雄性植株下降更多。此外，在Pb_{leaf}单独处理及Pb_{leaf}和干旱组合处理下，两性植株的PSⅡ有效光量子产量（$\Phi_{PSⅡ}$），尤其是雌性植株的$\Phi_{PSⅡ}$下降明显。$\Phi_{PSⅡ}$的下降可能是由于在光合作用的暗反应中，碳代谢能力的下降以及ATP和NADPH利用效率的下降。

暗适应叶片的PSⅡ潜在的光化学效率（F_v/F_m）是反映光抑制或光氧化对PSⅡ影响的一个很好的指示指标。Pb_{leaf}处理明显减少了两性植株尤其是雌性植株的潜在PSⅡ效率（F_v/F_m）［图6-3(b)］。对比之下，Pb_{soil}处理只略微减少了两性植株的潜在PSⅡ效率［图6-3(a)］。如果F_v/F_m大于0.8，PSⅡ的潜在效率则未被影响。在本研究中，所有处理的F_v/F_m均小于0.8，

尤其是在 Pb_{soil} 和干旱以及 Pb_{leaf} 和干旱组合处理中。值得注意的是，铅胁迫对 PSⅡ 潜在效率的影响可能取决于叶片的铅含量。在本研究中，在 Pb_{leaf} 处理下，雌雄植株叶片的铅含量很高，其 PSⅡ 潜在效率下降更多。而在 Pb_{soil} 处理下，雌雄植株叶片的铅含量很低，其 PSⅡ 潜在效率下降更少 [表 6-3、图 6-3(b)、图 6-3(a)]。

6.2.4　青杨雌雄植株对铅的吸收、转移和修复潜力

尽管在不同研究中，植物对铅的吸收有很大的变化，总的来看，根部均明显吸收了大量的铅，并限制了铅向地上部分的运输。我们的研究结果表明，雌雄植株的铅吸收和运输存在明显的性别间差异。在 Pb_{soil} 单独处理及 Pb_{soil} 和干旱组合处理中，两性植株均在根部积累了大量的铅，而地上部分只积累了少量的铅（表 6-3）。雌雄植株根部和叶片的铅吸收没有明显的性别间差异，而茎部的铅吸收则存在显著的性别间差异。在 Pb_{soil} 单独处理及 Pb_{soil} 和干旱组合处理中，雌性植株茎铅含量分别是雄性植株茎铅含量的 18.4 倍和 11.5 倍（表 6-3）。茎铅含量明显的性别间差异可能涉及两个特性：首先，相比于雌性植株，雄性植株可能有一个更加有效的内膜系统（如内皮层），因而限制了铅从根部向地上部分的运输。其次，雄性植株和雌性植株可能有不同的木质部解剖学特性（如不同的导管和管胞结构），这种差异影响了重金属在茎部的运输。相比之下，在 Pb_{leaf} 处理中，铅主要被叶片吸收，并且雌叶比雄叶吸收了更多的铅，这可能是因为雌叶和雄叶具有不同的叶特性，如不同的叶面积、表面蜡质、绒毛、气孔数量、上皮细胞形态特征及化学特性。此外，干旱明显减少了两性植株经由叶片的铅吸收，这可能是因为在干旱胁迫条件下，叶面积明显减少，气孔部分关闭，进而直接影响了铅经叶片的吸收。

第6章 青杨雌雄植株对干旱胁迫、铅胁迫及交互胁迫的响应差异

为了确定雌雄植株对铅污染土壤的植物修复潜力，植物修复相关参数，即生物浓度因子（BCF）、转移系数（T_f）和容忍系数（T_i）被分析。植物对重金属的容忍特性能够影响其对重金属的修复过程。基于整株植物的干重，容忍系数揭示在 Pb_{soil} 单独处理条件下，雌雄植株对铅容忍没有明显的差异。而在 Pb_{soil} 和干旱组合处理条件下，雄性植株的铅容忍性明显比雌性植株高（表6-4）。基于 Lux 等（2004）提出的标准，在 Pb_{soil} 单独处理条件下，雌雄可以被定义为高容忍的植物（$T_i > 60$），而在 Pb_{soil} 和干旱组合处理条件下，雄性植株表现为中度容忍（$35 < T_i < 60$），而雌性植株表现为低容忍（$T_i < 35$）。另外，生物浓度因子分析为判断雌雄植株从污染土壤提取重金属的能力提供了进一步的信息。在本研究中，雌雄植株表现出了不同的根 BCF 和地上部分 BCF。相比于雄性植株，雌性植株从土壤中富集重金属的能力更强。此外，通过计算转移因子（T_f）也可以进一步分析雌雄植株地上部分富集金属的能力。在本研究中，Pb_{soil} 单独处理条件下，雌雄植株的 T_f 没有明显的性别间差异，而在 Pb_{soil} 和干旱组合处理条件下，雌性植株比雄性植株具有更高的 T_f（表6-4）。综合这三个参数的分析，尽管在 Pb_{soil} 和干旱组合处理条件下，相比于雄性植株，雌性植株具有更高的 T_f 和 BCF，雌性植株仍然不适合植物提取，因为其对重金属的容忍性很差（$T_i < 35$）。总之，综合雌雄植株的铅积累、分布和植物修复能力相关系数的研究，在铅污染样地的植物修复策略中，雌雄植株可以用于实现不同的目的：一方面，研究表明雌性青杨是一种极有希望的重金属提取候选植物，因为其地上部分具有较高的重金属富集能力，但是其对重金属和干旱的容忍性仍需提高，以改善其在污染样地的存活率和生长状况。对比之下，雄性青杨更适合于植物固定技术，因为其根部富集铅的能力很强，并且对铅和干旱的容忍性也较高，因此可以有效地用于重金属固定，防

止重金属流失。

6.2.5 青杨雌雄植株叶片细胞超微结构的变化

在 Pb_{leaf} 处理的叶片中，较高的铅含量严重损伤了两性植株尤其是雌性植株的叶绿体、线粒体和细胞核［图 6-6(j)］，并直接导致其光合作用下降。但是对比来看，在 Pb_{soil} 处理中，铅元素主要在根部累积，只有少量的铅元素通过根吸收并向上运输到达叶绿体。因此，铅对光合作用器官的直接影响被大大限制。与 Pb_{soil} 处理相比，在 Pb_{leaf} 处理中，雌性植株的叶绿体又小又少，且相当畸形，内膜系统排列混乱［图 6-6(j)、图 6-6(l)］。叶绿体数量的减少可能直接导致雌性植株叶绿素含量的大幅下降［图 6-2(f)］。Baryla 等（2001）的研究也发现重金属严重抑制了油菜叶绿体的复制。另外，相比于 Pb_{soil} 单独处理及 Pb_{soil} 和干旱组合处理，在 Pb_{leaf} 单独处理及 Pb_{leaf} 和干旱组合处理中，叶绿体基质中分布有数量更多以及体积更大的质体小球［图 6-6(j)、图 6-6(l)］。质体小球是叶绿体衰老的一个表征，其含有类囊体降解的脂肪滴。在铜胁迫处理的玉米叶片中也观察到了类似的叶绿体变化。此外，在 Pb_{leaf} 和干旱组合处理条件下，叶绿体超微结构的变化（如类囊体肿胀和排列紊乱）也是铅诱导的植物细胞衰老信号之一［图 6-6(l)］。

§6.3 小结

我们的研究结果表明，Pb_{leaf} 单独处理和组合处理（Pb_{leaf}＋干旱以及 Pb_{soil}＋干旱）明显影响了青杨雌雄植株的干物质积累、光合作用速率、长期水分使用效率、叶绿素荧光参数以及细胞超

第6章 青杨雌雄植株对干旱胁迫、铅胁迫及交互胁迫的响应差异

微结构,并且对雌性植株的负影响大于雄性植株。但是 Pb_{soil} 单独处理几乎没有影响两性植株的形态学和生理学特性。通过对两性植株在 Pb_{soil} 单独处理及 Pb_{soil} 和干旱组合处理条件下的生物浓度因子(BCF)、容忍系数(T_i)和转移系数(T_f)的分析表明,雌雄植株适用于不同的植物修复策略,雄性植株有希望被开发为植物固定候选植物,雌性植株有希望被开发为植物提取候选植物。

第7章 青杨雌雄交互嫁接植株对干旱胁迫的响应差异

§7.1 主要研究结果

7.1.1 不同水分处理下青杨雌雄交互嫁接植株的干物质累积和分配差异

干旱明显减少了所有嫁接组合的叶、茎和根的干物质累积[图7-1(a)]。在水分亏缺的条件下,与自嫁接的雄性植株(M/M)或交互嫁接的个体(M/F和F/M)相比,自嫁接的雌性植株(F/F)积累了较少的干物质[图7-1(a)]。有趣的是,在干旱胁迫条件下,相比于F/F和M/M,F/M嫁接组合积累了更多的干物质[图7-1(a)]。总的来看,干旱胁迫条件下,雄根组合植株比雌根组合植株积累的干物质更多。在水分充足的条件下,4个嫁接组合的干物质累积量明显不同,相比于其他3个嫁接组合,F/F累积了更多的干物质。更加有趣的是,在水分充足的条件下,相比于F/F,F/M累积了更少的干物质;但相比于M/M,其累积的干物质却明显更多[图7-1(a)]。很显然在水分充足的条件下,雌雄植株个体间的生长差异主要取决于植株地上部分,与植株地下部分的基因型关系不大,即相比于地上部分为雄性的植株，地上部分为雌性的植株累积了更多的干物质。此外，干旱胁迫还提高了所有嫁接组合的根茎比［图7-1(b)］。在水分亏缺的条件下，相比于雌根组合植株，雄根组合植株的根茎比增加更多。因此，在水分亏缺的条件下，雄根组合分配了较多的干物质于根部，而相比之下，雌根组合则分配了较少的干物质于根部［图7-1(b)］。

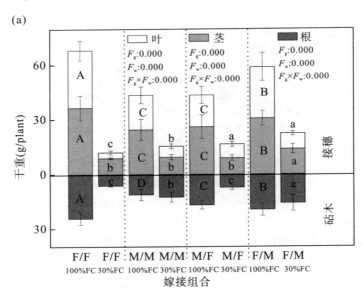

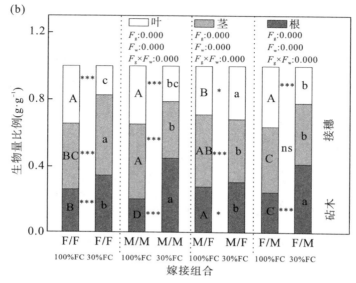

图 7-1 不同嫁接组合植株的生物量累积和分配

注：1. (a) 植物不同部位的干重，数值为平均值±标准误差（$n=10$）；(b) 干重在植物不同部位的分配比例（不包括残留的 cutting）。

2. 每个柱上不同的大写字母和小写字母分别表示在水分充足和水分亏缺条件下，嫁接组合间在 $P<0.05$ 水平上差异显著（Tukey 检验）。* 表示在嫁接组合内干旱处理和对照之间差异显著（独立样本 t 检验）。*，$0.01<P<0.05$；**，$0.001<P<0.01$；***，$P\leqslant0.001$；ns，差异不显著。

3. 基因型符号为接穗/砧木。

4. F_g，嫁接类型的影响；F_w，干旱处理的影响；$F_g \times F_w$，嫁接类型和干旱交互处理的影响。

7.1.2 不同水分处理下青杨雌雄交互嫁接植株的气体交换参数差异

水分亏缺的条件下，所有嫁接组合的 P_n 均明显下降，相比于雌根组合，雄根组合的 P_n 明显更高。而在水分充足的条件下，

第7章 青杨雌雄交互嫁接植株对干旱胁迫的响应差异

相比于地上部分为雄性植株的组合，地上部分为雌性植株的组合则具有明显更高的 P_n（表7-1）。另外，在水分充足条件下，所有嫁接组合的 C_i/C_a 却没有明显的差异，其值仅在 0.67~0.78 之间略微变动。而在水分亏缺的条件下，相比 F/F，M/M 具有明显更高的 C_i/C_a，但在其他嫁接组合间，C_i/C_a 并无明显差异。

表7-1 水分充足和亏缺条件下4个嫁接组合的净光合作用速率（P_n）以及细胞间 CO_2 和环境 CO_2 浓度比（C_i/C_a）

田间持水量/%	S/R	P_n/($\mu mol \cdot m^{-2} \cdot s^{-1}$)	C_i/C_a
100	F/F	18.01±0.35A	0.67±0.01A
	M/M	11.10±0.41D	0.74±0.05A
	M/F	14.06±0.24C	0.71±0.02A
	F/M	16.33±0.49B	0.78±0.02A
变异源（平方和）	组间（$df_1=3$）	134.513	0.037
	组内（$df_2=16$）	11.726	0.069
30	F/F	2.57±0.11d	0.41±0.02b
	M/M	6.41±0.10b	0.50±0.02a
	M/F	5.79±0.19c	0.48±0.02ab
	F/M	8.68±0.16a	0.48±0.02ab
变异源（平方和）	组间（$df_1=3$）	9.513	0.041
	组内（$df_2=16$）	0.338	0.153
P	F_g	***	**
	F_w	***	***
	$F_g \times F_w$	***	ns

注：1. 每个数值均是平均值 ± 标准误（$n=5$）。每列不同的大写字母和小写字母分别表示水分充足和水分亏缺条件下，嫁接组合间在 $P<$

0.05 水平上差异显著（Tukey 检验）。

2. 基因型符号为接穗/砧木。

3. F_g，嫁接类型的影响；F_w，干旱处理的影响；$F_g \times F_w$，嫁接类型和干旱交互处理的影响。

7.1.3 不同水分处理下雌雄交互嫁接植株的用水效率差异

在水分充足的条件下，不同嫁接组合间的 Δ 差异相当小，仅在 24.57‰~24.00‰ 之间小幅度变动（图 7-2）。干旱胁迫明显减少了 4 个嫁接组合的 Δ。相比之下，在水分亏缺的条件下，不同嫁接组合间的 Δ 差异较大，在 23.29‰~22.43‰ 之间变化。在水分亏缺和水分充足的条件下，相比于雄根组合，雌根组合植株 Δ 明显更高（图 7-2）。值得注意的是，尽管在本研究中 ^{13}C 的丰度非常高（超过 1‰），但相比于其他研究，本研究的 Δ 明显偏高。这可能是由于 30% 田间持水量的干旱处理对植株的胁迫强度较大；也可能是因为在密闭温室环境中，呼吸产生的 CO_2 以及消耗的 CO_2 大量再循环。

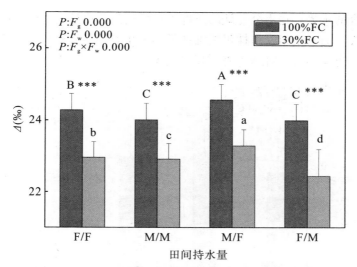

图 7-2 水分充足和亏缺条件下 4 个嫁接组合的碳同位素分馏（Δ）分析

注：1. 每个数值均是平均值±标准误差（$n = 5$）。

2. 每个柱上不同的大写字母和小写字母分别表示在水分充足和水分亏缺条件下，嫁接组合间在 $P<0.05$ 水平上差异显著（Tukey 检验）。*表示在嫁接组合内干旱处理和对照之间差异显著（独立样本 t 检验）。*，$0.01<P<0.05$；**，$0.001<P<0.01$；***，$P \leqslant 0.001$；ns，差异不显著。

3. 基因型符号为接穗/砧木。

4. F_g，嫁接类型的影响；F_w，干旱处理的影响；$F_g \times F_w$，嫁接类型和干旱交互处理的影响。

7.1.4 不同水分处理下青杨雌雄交互嫁接植株的黎明前叶水势差异

在水分充足的条件下，与地上部分为雌性的嫁接组合相比，地上部分为雄性的嫁接组合有较高的黎明前叶水势，其值在 $-0.81 \sim -0.76$ MPa 变化（图 7-3）。干旱明显降低了所有嫁接组合的水势。在水分亏缺的条件下，雌根组合植株的水势下降非常大，而雄根组合植株仍旧维持了一个相对较高的水势，其值在 $-1.82 \sim -1.52$ MPa 变化（图 7-3）。

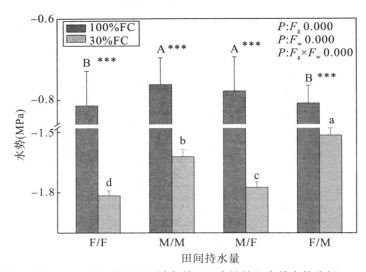

图 7-3 水分充足和亏缺条件下 4 个嫁接组合的水势分析

注：1. 每个数值均是平均值±标准误差（$n=5$）。

2. 每个柱上不同的大写字母和小写字母分别表示在水分充足和水分亏缺条件下，嫁接组合间在 $P<0.05$ 水平上差异显著（Tukey 检验）。＊号表示在嫁接组合内干旱处理和对照之间差异显著（独立样本 t 检验）。＊，$0.01<P<0.05$；＊＊，$0.001<P<0.01$；＊＊＊，$P \leqslant 0.001$；ns，差异不显著。

3. 基因型符号为接穗/砧木。

4. F_g，嫁接类型的影响；F_w，干旱处理的影响；$F_g \times F_w$，嫁接类型和干旱交互处理的影响。

7.1.5 不同水分处理下青杨雌雄交互嫁接植株叶肉细胞超微结构的差异

在本研究中，我们利用透射电子显微镜观察了嫁接组合叶肉细胞超微结构的变化。研究发现，在水分充足的条件下，所有嫁接组合叶肉细胞的细胞膜和细胞壁均平滑、清晰和连续，颗粒状的细胞质密集填充于各种细胞器，染色质均匀分布于核基质中，叶绿体呈现典型的透镜形，类囊体膜排列规则，线粒体显示出清晰的嵴［图7-4(e)］。但是在水分亏缺的条件下，所有嫁接组合的细胞显示出不同的超微结构变化。在水分亏缺的条件下，与对照组相比，所有处理植株的叶绿体均含有大量的质体小球［图7-4(a)~图7-4(d)］，其中F/M组合的叶绿体略微肿胀，但是线粒体结构正常，细胞质基质均匀［图7-4(b)］。而M/M组合的叶绿体和线粒体均略微肿胀，线粒体的数量增加并聚集在一起［图7-4(d)］。在M/F组合中，一小部分叶绿体和线粒体出现肿胀并略微变形，染色质浓缩，核仁消失，核膜降解［图7-4(a)］。对比之下，干旱胁迫严重破坏了F/F组合的细胞结构，其叶绿体和线粒体明显肿胀，线粒体嵴消失并且空泡化，此外，细胞膜也被严重破坏［图7-4(c)］。

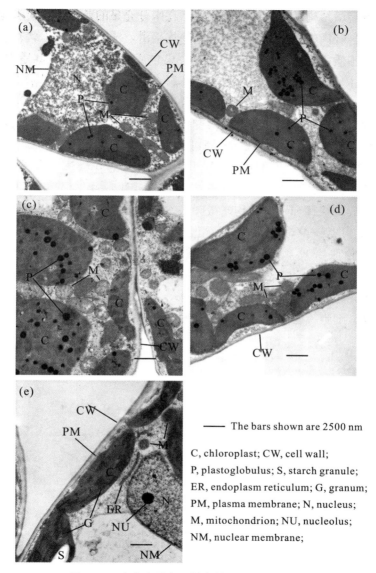

图 7-4 水分充足和亏缺条件下 4 个嫁接组合叶肉细胞的透射电子显微镜观察

注：(a) 水分亏缺条件下，雌性砧木＋雄性接穗（M/F）组合；(b) 水分亏缺条件下，雄性砧木＋雌性接穗（F/M）组合；(c) 水分亏缺条件下，雌性砧木＋雌性接穗（F/F）组合；(d) 水分亏缺条件下，雄性砧木＋雄性接穗（M/M）组合；(e) 水分充足条件下，雌性砧木＋雌性接穗（F/F）组合。水分充足条件下，其他3个嫁接组合显示出类似的细胞超微结构，因此图片未列出。

§7.2 讨论

7.2.1 嫁接组合对干旱胁迫的生理学响应

在全世界范围内，干旱是限制植物生长和生态系统生产力的最重要因素之一。植物对干旱的响应非常复杂，涉及很多方面，如干旱胁迫的感知和信号传导、生长和生物量分配模式的改变、水分状态的动态平衡、气孔导度和CO_2同化的下降、渗透调节以及脱毒过程等。我们先前的研究表明，在干旱胁迫条件下，相比于雌性植株，雄性植株的干物质累积能力和光合作用能力更强，并且具有较高的长期水分利用效率。而在本书研究中，相比于F/F，M/M具有更高的干物质累积能力、光合作用能力和长期水分利用效率，与先前研究相比，自嫁接并没有影响雌雄植株对干旱的生理学响应。但是当雌性接穗嫁接到雄性砧木上（F/M），雌性接穗（F/M）比雄性接穗（M/M和M/F）显示出更高的干物质累积能力、光合作用能力和长期水分利用效率。这一现象充分表明青杨雄性植株地下部分显著增强了雌性植株地上部分的干旱容忍性。

植物的干物质累积取决于CO_2同化的速率和持续时间、光合产物分配到叶面积的比例以及根茎比。在本研究中，在水分亏缺的条件下，相比于 F/F，F/M 累积了更多的干物质；而相比于 M/F，M/M 的干物质累积更多。这可能归咎于两种原因：一是可能在干旱处理前，迅速生长的雌性接穗提供了更多的碳水化合物供雄根生长发育，在之后的水分胁迫阶段，发达的雄根吸收了大量的水并转移了较多的水到植株的地上部分，提高了植株的抗旱性。如图 7-1(a) 所示，在水分充足的条件下，F/M 的根系统较发达，在干旱条件下，其干物质累积较多[图 7-1(a)]。二是可能在干旱条件下，雄根的水分吸收能力更强（水分吸收能力取决于根的水压传导特性），从而确保雌性接穗在干旱胁迫下能迅速生长。这两个解释均被目前的研究数据所支持[图 7-1(a)]。

碳同位素分馏（Δ）作为一种工具，能有效评估 C_3 植物的长期用水效率（WUE）。根据理论推导和无数实验证据，WUE 和 Δ 之间存在负线性关系。在本研究中，相比于雌根组合，雄根组合的 Δ 更小，这表明雄根组合的用水效率更高。然而值得注意的是，碳同位素分馏（图 7-2）和气体交换结果（C_i/C_a，表 7-1）之间存在明显的分歧，表 7-1 所显示的 C_i/C_a 与 Δ 并不一致。与 Δ 相比较，雄根组合的 C_i/C_a 相对较高。碳同位素分馏数据与气体交换数据之间的分歧可能主要是因为这两个数据的属性不一样：二者在时间尺度和信号滞留方面存在差异，从而对季节和环境波动的响应不一致。具体来说，碳同位素分馏分析能够提供不同时间尺度的整合，其取决于所分析植物碳的比例。在本实验中，碳同位素分馏主要代表了完全发育叶片的结构性碳。叶片的生长主要是在嫁接建立期间（即 2011 年 3 月到 5 月）完成的。只有一小部分碳是在干旱处理开始后（即 2011 年 6 月到 7 月）被固定的。相比之下，C_i/C_a 只反映了所测叶片的瞬时状态。此外，叶肉导度（g_m）可能也会影响 Δ 和 WUE 之间的负线性关系。

嫁接组合间水分吸收能力的变化［基于图7-1(a)推测］也非常有趣。通常根对水分的吸收取决于根水压传导特性（L_p）。研究发现，根水压传导能力非常容易变化，且叶水势（水分状态）存在一定的关系。在本研究中，在水分亏缺的条件下，雄根组合比雌根组合维持了相对较高的黎明前叶水势（图7-3），这种差异很可能是不同嫁接组合具有不同水分吸收能力的一个暗示，如Else等（2001）、Reich和Hinckley（1989）的研究表明，黎明前叶水势与根水压传导能力正相关。尽管如此，仍需开展进一步的研究，以提供不同嫁接组合根水压传导特性变化的确凿证据。

在超微结构水平，水分亏缺处理致使所有嫁接组合的叶绿体肿胀，并且在叶绿体基质中，质体小球数量和大小均明显增加［图7-4(a)、图7-4(b)、图7-4(c)、图7-4(d)］。相比于雄根组合叶绿体，在雌根组合叶绿体中，质体小球的大小和密度明显更高，这种差异在低放大倍数下尤其明显（数据未显示）。研究发现，质体小球是包含脂的结构并带有低水平的蛋白质。在类囊体形成过程中，质体小球的数量减少；而在叶绿体衰老的过程中，伴随着类囊体膜的退化，其数量增加。总的来看，相比于雄根组合，在雌根组合中，叶绿体肿胀更加明显。由于光合作用发生在叶绿体的类囊体膜上，因此，干旱胁迫下4种嫁接组合超微结构的差异为其在干旱条件下呈现出的不同光合作用能力提供了直接的细胞学证据。

7.2.2 青杨雌雄植株对干旱的响应差异主要取决于根

根部和地上部分内在的相互作用使得某一个性状（如对干旱的响应）很难确定，这是由根部表达的基因和地上部分表达的基因控制的，还是由整个植株表达的基因所控制的，还不清晰。而通过嫁接，交换地上部分和地下部分的基因型，可以获得这样的信息。在本研究中,我们通过青杨雌雄植株之间的交互嫁接,确定了在水分充足和水分亏缺的条件下,青杨雌雄植株地上部分和地下部分的相对重要性。研究发现,青杨雄性植株的根能明显改善雌性植株的干旱胁迫抗性。具体表现为:在水分亏缺的条件下,相比于雌根组合,雄根组合的生长抑制和光合作用速率下降更少、水势和用水效率更高以及细胞器受到的损伤更小。尽管目前没有性相关的根生理学信息,但是根据本研究我们可以推断,青杨雌雄植株一定存在明显不同的根生理学特性。因此,进一步开展性相关根生理学差异的研究将为揭示雌雄异株植物性相关生理学差异机制提供新的线索。根在性相关干旱胁迫响应中的作用可能是长期自然选择的结果,雌雄植株间不同的生殖成本对根的生长和活性产生了不同的影响。

7.2.3 嫁接提高了雌性青杨的水分亏缺容忍性

杨树是生长迅速的树种之一,它们较高的生产力与大量的水分需求密切相关,所以其生产力主要取决于水分的可利用性。水分亏缺容忍性这一概念应用于栽培树种(如杨树),具体指在中度水分亏缺的条件下,树木维持较高生物量的能力。在本研究中我们发现,在水分亏缺的条件下,相比于其他 3 种嫁接组合,F/M 组合的生物量累积最多 [图 7-1(a)]。因此,将青杨雌性植株接穗嫁接到青杨雄性植株砧木上,可以有效改善青杨雌性植株的水分亏缺容忍性,该方法可能是一种行之有效的育种途径。此外研究发现,青杨雌雄植株间的嫁接相容性非常好,嫁接植株存活率为 100%,并且生长良好。

§7.3 小结

目前的研究表明，青杨雌雄植株对干旱胁迫的响应差异主要取决于植株的地下部分，而不依赖于植株的地上部分。然而在水分充足的条件下，青杨雌雄植株的生长差异主要由地上部分驱动，在很大程度上不依赖于根部。此外，将雌性接穗嫁接到雄性砧木上是提高青杨雌性植株水分亏缺容忍性的一种行之有效的手段。

主要结论

本研究以青杨雌雄植株为模式植物,研究了不同性别青杨一年生扦插幼苗对干旱、铅及其交互作用的不同响应,揭示了青杨雌雄植株在生物量累积和分配、气体交换参数、叶绿素荧光参数、铅累积和分布、植物修复相关参数以及叶肉细胞超微结构的差异。另外,巧妙利用嫁接手段,研究了青杨雌雄植株交互嫁接组合对干旱的响应差异,揭示了青杨雌雄植株地上部分和地下部分在干旱差异响应中的作用。研究成果丰富了雌雄异株木本植物对逆境胁迫差异响应的生理生态研究,并且为揭示青杨雌雄植株的干旱差异响应机制提供了重要线索,同时为林业育种和重金属污染样地的植物修复提供理论依据和重要参考。主要结论如下:

(1)青杨雌雄植株对干旱胁迫、铅胁迫及其交互作用的响应差异。

叶面喷洒铅处理(Pb_{leaf})、叶面喷洒铅和干旱组合处理(Pb_{leaf}+干旱)以及土壤添加铅和干旱组合处理(Pb_{soil}+干旱)明显影响了青杨雌雄植株的干物质积累、气体交换参数、长期水分使用效率、叶绿素荧光参数以及细胞超微结构,并且相比于雄性植株,雌性植株受到的不利影响更大。但是土壤添加铅处理(Pb_{soil})几乎没有对雌性植株和雄性植株的形态学和生理学特性产生影响。通过对雌雄植株在Pb_{soil}及Pb_{soil}+干旱条件下的生物浓度因子(BCF)、容忍系数(T_i)和转移系数(T_f)及铅含量和分布的分析表明，雌性植株和雄性植株分别适用于不同的植物修复策略，雄性植株有望被开发为植物固定候选树种，而雌性植株有望被开发为植物提取候选树种。

（2）青杨雌雄交互嫁接植株对干旱的响应差异。

在水分亏缺的条件下，雄根组合（F/M 和 M/M）比雌根组合（F/F 和 M/F）有更高的干物质积累（DMA）、根茎比（R/S）、净光合作用速率（P_n）、黎明前叶水势（ψ_{pd}）以及长期水分利用效率（WUE），并且其细胞超微结构的损伤较轻。但在水分充足的条件下，雌接穗组合（F/F 和 F/M）比雄接穗组合（M/M 和 M/F）有较高的 DMA、R/S 和 P_n，以及较低的 WUE 和 ψ_{pd}。这些结果表明，青杨雌雄植株对水分亏缺的敏感性主要由定位于根部的机制所驱动，而不主要依赖于地上部分的基因型。然而在水分充足的条件下，青杨雌雄植株的生长差异主要由地上部分驱动，在很大程度上并不依赖于根系。此外，将雌性接穗嫁接到雄性砧木上是提高青杨雌性植株水分亏缺容忍性的一种非常有效的手段。

参考文献

[1] Abadelhafeez A T, Harssema H, Verkerk K. Effects of air temperature, soil temperature and soil moisture on growth and development of tomato itself and grafted on its own and eggplant rootstock [J]. Scientia Horticulturae, 1975, 3: 65−73.

[2] Abdelmageed A H A, Gruda N. Influence of grafting on growth, development and some physiological parameters of tomatoes under controlled heat stress conditions [J]. European Journal of Horticultural Science, 2009, 74: 16−20.

[3] Abeles F B, Morgan P W, Saltveit Jr M E. Ethylene in plant biology [M]. New York: Academic Press, 2012.

[4] Aganchich B, Wahbi S, Loreto F, et al. Partial root zone drying: regulation of photosynthetic limitations and antioxidant enzymatic activities in young olive (*Olea europaea*) saplings [J]. Tree Physiology, 2009, 29: 685−696.

[5] Ågren J, Danell K, Elmqvist T, et al. Sexual dimorphism and biotic interactions. Gender and sexual dimorphism in flowering plants [M]. Heidelberg: Springer-Verlag, 1999.

[6] Ahn S J, Im Y J, Chung G C, et al. Physiological

responses of grafted-cucumber leaves and rootstock roots affected by low root temperature [J]. Scientia Horticulturae, 1999, 81: 397—408.

[7] Albacete A, Ghanem M E, Martinez-Andujar C, et al. Hormonal changes in relation to biomass partitioning and shoot growth impairment in salinized tomato (*Solanum lycopersicum* L.) plants [J]. Journal of Experimental Botany, 2008, 59: 4119—4131.

[8] Albacete A, Martinez-Andujar C, Ghanem M E, et al. Rootstock-mediated changes in xylem ionic and hormonal status are correlated with delayed leaf senescence, and increased leaf area and crop productivity in salinized tomato [J]. Plant, Cell & Environment, 2009, 32: 928—938.

[9] Ali I A, Kafkafi U, Yamaguchi I, et al. Effects of low root temperature on sap flow rate, soluble carbohydrates, nitrate contents and on cytokinin and gibberellin levels in root xylem exudate of sand-grown tomato [J]. Journal of Plant Nutrition, 1996, 19: 619—634.

[10] Alsina M M, Smart D R, Bauerle T, et al. Seasonal changes of whole root conductance by a drought-tolerant grape root system [J]. Journal of Experimental Botany, 2011, 62: 99—109.

[11] Alvarez-Cansino L, Zunzunegui M, Barradas M C D, et al. Gender-specific costs of reproduction on vegetative growth and physiological performance in the dioecious shrub *Corema album* [J]. Annals of Botany, 2010, 106: 989—998.

[12] Arias J A, Peralta-Videa J R, Ellzey J T, et al. Effects

of Glomus deserticola inoculation on Prosopis: enhancing chromium and lead uptake and ranslocation as confirmed by X-ray mapping, ICP-OES and TEM techniques [J]. Environmental and Experimental Botany, 2010, 68: 139−148.

[13] Aroca R, Ferrante A, Vernieri P, et al. Drought, abscisic acid and transpirationrate effects on the regulation of PIP aquaporin gene expression and abundance in *Phaseolus vulagaris* plants [J]. Annals of Botany, 2006, 98: 1301−1310.

[14] Aroca R, Porcel R, Ruiz-Lozano J M. How does arbuscular mycorrhizal symbiosis regulate root hydraulic properties and plasma membrane aquaporins in *Phaseolus vulgaris* under drought, cold or salinity stresses? [J]. New Phytologist, 2007, 173: 808−816.

[15] Aroca R, Porcel R, Ruiz-Lozano J M. Regulation of root water uptake under abiotic stress conditions [J]. Journal of Experimental Botany, 2012, 63: 42−57.

[16] Arshad M, Silvestre J, Pinelli E, et al. A field study of lead phytoextraction by various scented *Pelargonium cultivars* [J]. Chemosphere, 2008, 71: 2187−2192.

[17] Asins M J, Bolarín M C, Pérez-Alfocea F, et al. Genetic analysis of physiological components of salt tolerance conferred by Solanum rootstocks. What is the rootstock doing for the scion? [J]. Theoretical and Applied Genetics, 2010, 121: 105−115.

[18] Atici Ö, Aǧar G, Battal P. Changes in phytohormone contents in chickpea seeds germinating under lead or zinc

stress [J]. Biologia Plantarum, 2005, 49: 215-222.

[19] Atkin R K, Barton G E, Robinson D K. Effect of root-growing temperature on growth substances in xylem exudate of *Zea mays* [J]. Journal of Experimental Botany, 1973, 24: 475-487.

[20] ATSDR. Agency for toxic substances and disease registry [EB/OL]. [2021-08-02]. http://www.atsdr.cdc.gov/.

[21] Barceló J, Poschenrieder C. Plant water relations as affected by heavy metal stress: a review [J]. Journal of Plant Nutrition, 1990, 13: 1-37.

[22] Baryla A, Carrier P, Franck F, et al. Leaf chlorosis in oilseed rape plants (*Brassica napus*) grown on cadmium-polluted soil: causes and consequences for photosynthesis and growth [J]. Planta, 2001, 212: 696-709.

[23] Becerril J M, Gonzales-Marua C, Munoz-Rueda A, et al. Changes induced by cadmium and lead in gas exchange and water relations of clover and leucerne [J]. Plant Physiology and Biochemistry, 1989, 27: 913-918.

[24] Bi X, Feng X, Yang Y, et al. Quantitative assessment of cadmium emission from zinc smelting and its influences on the surface soils and mosses in Hezhang County, Southwestern China [J]. Atmospheric Environment, 2006, 40: 4228-4233.

[25] Björkman O, Demmig B. Photon yield of O_2 evolution and chlorophyll fluorescence characteristics at 77K among vascular plants [J]. Planta, 1987, 170: 89-504.

[26] Bloom A J, Zwieniecki M A, Passioura J B, et al. Water relations under root chilling in a sensitive and tolerant

tomato species [J]. Plant, Cell & Environment, 2004, 27: 971−979.

[27] Bluskov S, Arocena J M, Omotoso O O, et al. Uptake, distribution, and speciation of chromium in *Brassica juncea* [J]. International Journal of Phytoremediation, 2005, 7: 153−165.

[28] Bota J, Medrano H, Flexas J. Is photosynthesis limited by decreased Rubisco activity and RuBP content under progressive water stress? [J]. New Phytologist, 2004, 162: 671−681.

[29] Brugnoli E, Farquhar G D. Photosynthetic fractionation of carbon isotopes [M] //Leegood R C, Sharkey T D, von Caemmerer S. Photosynthesis: physiology and metabolism, Advances in photosynthesis. Vol. 9. The Netherlands: Kluwer Academic Publishers, 2003: 9−434.

[30] Brunet J, Varrault G, Zuily-Fodil Y, et al. Accumulation of lead in the roots of grass pea (*Lathyrus sativus* L.) plants triggers systemic variation in gene expression in the shoots [J]. Chemosphere, 2009, 77: 1113−1120.

[31] Bugbee B, White J W. Tomato growth as affected by root-zone temperature and the addition of gibberellic acid and kinetin to nutrient solution [J]. Journal of the American Society for Horticultural Science, 1984, 109: 121−125.

[32] Bulder H A M, van Hasselt P R, Kuiper P J C. The effect of temperature on early growth of cucumber genotypes differing in genetic adaptation to lowenergy

conditions [J]. Scientia Horticulturae, 1987, 31: 53−60.

[33] Centritto M, Lauteri M, Monteverdi M C, et al. Leaf gas exchange, carbon isotope discrimination, and grain yield in contrasting rice genotypes subjected to water deficits during the reproductive stage [J]. Journal of Experimental Botany, 2009, 60: 2325−2339.

[34] Centritto M, Loreto F, Chartzoulakis K. The use of low [CO_2] to estimate diffusional and non-diffusional limitation of photosynthetic capacity of salt-stressed olive saplings [J]. Plant, Cell & Environment, 2003, 26: 585−594.

[35] Chatterjee C, Dube B K, Sinha P, et al. Detrimental effects of lead phytotoxicity on growth, yield, and metabolism of rice [J]. Communications in Soil Science and Plant Analysis, 2004, 35: 255−265.

[36] Chaves M M, Flexas J, Pinheiro C. Photosynthesis under drought and salt stress: regulation mechanisms from whole plant to cell [J]. Annals of Botany, 2009, 103: 551−560.

[37] Chaves M M, Maroco J P, Pereira J S. Understanding plant responses to drought: from genes to the whole plant [J]. Functional Plant Biology, 2003, 30: 239−264.

[38] Choi K J, Chung G C, Ahn S J. Effect of root-zone temperature on the mineral composition of xylem sap and plasmamembrane K^+-Mg^{2+}-ATPase activity of grafted-cucumber and figleaf gourd root systems [J]. Plant and Cell Physiology, 1995, 36: 639−643.

[39] Clearwater M J, Lowe R G, Hofstee B J, et al.

Hydraulic conductance and rootstock effects in grafted vines of kiwifruit [J]. Journal of Experimental Botany, 2004, 55: 1371−1381.

[40] Colla G, Rouphael Y, Leonardi C, et al. Role of grafting in vegetable crops grown under saline conditions [J]. Scientia Horticulturae, 2010, 127: 147−155.

[41] Correia O, Díaz Barradas M C. Ecophysiological differences between male and female plants of *Pistacia lentiscus* L. [J]. Plant Ecology, 2000, 149: 31−142.

[42] Criddle R S, Smith B N, Hansen L D. A respiration based description of plant growth rate responses to temperature [J]. Planta, 1997, 201: 441−445.

[43] Daie J, Campbell W F. Response of tomato plants to stressful temperatures. Increase in abscicic acid concentrations [J]. Plant Physiology, 1981, 67: 26−29.

[44] Darwin C. On the origin of species by means of natural selection, or the preservation of favoured races in the struggle for life [J]. The British and Foreign Medico-chirurgical Review, 1860, 25: 367−404.

[45] Darwin C. The different forms of flowers on plants of the same species [M]. London: John Murray, 1877.

[46] Dawson T E, Bliss L C. Patterns of water use and the tissue water relations in the dioecious shrub, *Salix arctica*: the physiological basis for habitat partitioning between the sexes [J]. Oecologia, 1989, 79: 332−343.

[47] Dawson T E, Bliss L C. Plants as mosaics: leaf-, ramet-, and gender-level variation in the physiology of the dwarf willow, Salix arctica [J]. Functional Ecology,

1993, 7: 293—304.

[48] Dawson T E, Ehleringer J R. Gender-specific physiology, carbon isotope discrimination, and habitat distribution in boxelder, *Acer negundo* [J]. Ecology, 1993, 74: 798—815.

[49] Delph L F, Galloway L F, Stanton M L. Sexual dimorphism in flower size [J]. The American Naturalist, 1996, 148: 299—320.

[50] Delph L F. Sexual dimorphism in life history. Gender and sexual dimorphism in flowering plants [M]. Heidelberg: Springer-Verlag, 1999.

[51] Demmig-Adams B, Adams W W. Photoprotection in an ecological context: the remarkable complexity of thermal energy dissipation [J]. New Phytologist, 2006, 172: 11—21.

[52] Dey S K, Dey J, Patra S, et al. Changes in the antioxidative enzyme activities and lipid peroxidation in wheat seedlings exposed to cadmium and lead stress [J]. Brazilian Journal of Plant Physiology, 2007, 19: 53—60.

[53] Di Baccio D, Tognetti R, Minnocci A, et al. Responses of the *Populus x euramericana* clone I -214 to excess zinc: carbon assimilation, structural modifications, metal distribution and cellular localization [J]. Environmental and Experimental Botany, 2009, 67: 153—163.

[54] Di Baccio D, Tognetti R, Sebastiani L, et al. Responses of *Populus deltoides* × *Populus nigra* (*Populus* × *euramericana*) clone I -214 to high zinc concentrations [J]. New Phytologist, 2000, 159: 443—452.

[55] Disante K B, Fuentes D, Cortina J. Sensitivity to zinc of Mediterranean woody species important for restoration [J]. Science of the Total Environment, 2010, 408: 2216−2225.

[56] Else M A, Coupland D, Dutton L, et al. Decreased root hydraulic conductivity reduces leaf water potential, initiates stomatal closure and slows leaf expansion in flooded plants of castor oil (*Ricinus communis*) despite diminished delivery of ABA from the roots to shoots in xylem sap [J]. Physiologia Plantarum, 2001, 111: 46−54.

[57] Elzbieta W, Miroslawa C. Lead-induced histological and ultrastructural changes in the leaves of soybean (*Glycine max* L.) [J]. Soil Science & Plant Nutrition, 2005, 51: 203−212.

[58] Farquhar G D, Ehleringer J R, Hubick K T. Carbon isotope discrimination and photosynthesis [J]. Annual Review of Plant Biology, 1989, 40: 503−537.

[59] Farquhar G D, O'leary M H, Berry J A. On the relationship between carbon isotope discrimination and the intercellular carbon dioxide concentration in leaves [J]. Functional Plant Biology, 1982, 9: 121−137.

[60] Feng X, Li G, Qiu G. A preliminary study on mercury contamination to the environment from artisanal zinc smelting using indigenous method in Hezhang County, Guizhou, China: Part 2. Mercury contaminations to soil and crop [J]. Science of the Total Environment, 2006, 368: 47−55.

[61] Flexas J, Bota J, Cifre J, et al. Understanding downregulation of photosynthesis under water stress: future prospects and searching for physiological tools for irrigation management [J]. Annals of Applied Biology, 2004, 144: 273−283.

[62] Flexas J, Bota J, Escalona J M, et al. Effects of drought on photosynthesis in grapevines under field conditions: an evaluation of stomatal and mesophyll limitations [J]. Functional Plant Biology, 2002, 29: 461−471.

[63] Flexas J, Bota J, Galmés J, et al. Keeping a positive carbon balance under adverse conditions: responses of photosynthesis and respiration to water stress [J]. Physiologia Plantarum, 2006, 127: 343−352.

[64] Flexas J, Bota J, Loreto F, et al. Diffusive and metabolic limitations to photosynthesis under drought and salinity in C_3 plants [J]. Plant Biology, 2004, 6: 269−279.

[65] Flexas J, Ribas-Carbo M, Bota J, et al. Decreased rubisco activity during water stress is not induced by decreased relative water content but related to conditions of low stomatal conductance and chloroplast CO_2 concentration [J]. New Phytologist, 2006, 172, 73−82.

[66] Flexas J, Ribas-Carbó M, Diaz-Espejo A, et al. Mesophyll conductance to CO_2: current knowledge and future prospects [J]. Plant, Cell & Environment, 2008, 31: 602−621.

[67] Flexas J, Ribas-Carbó M, Hanson D T, et al. Tobacco aquaporin NtAQP1 is involved in mesophyll conductance to

CO_2 in vivo [J]. The Plant Journal, 2006, 48: 427−439.

[68] Fox J F. Shoot demographic responses to manipulation of reproductive effort by bud removal in a willow [J]. Oikos, 1995, 72: 283−287.

[69] Gajewska E, Slaba M, Andrzejewska R, et al. Nickel-induced inhibition of wheat root growth is related to H_2O_2 production, but not to lipid peroxidation [J]. Plant Growth Regulation, 2006, 49: 95−103.

[70] Galmés J, Conesa M A, Ochogavia J M, et al. Physiological and morphological adaptations in relation to water use efficiency in Mediterranean accessions of *Solanum lycopersicum* [J]. Plant, Cell & Environment, 2011, 34: 245−260.

[71] Galmés J, Medrano H, Flexas J. Photosynthetic limitations in response to water stress and recovery in Mediterranean plants with different growth forms [J]. New Phytologist, 2007, 175: 81−93.

[72] Galmés J, Ribas-Carbo M, Medrano H, et al. Rubisco activity in Mediterranean species is regulated by the chloroplastic CO_2 concentration under water stress [J]. Journal of Experimental Botany, 2011, 62: 653−665.

[73] Gao J J, Qin A G, Yu X C. Effect of grafting on cucumber leaf SOD and CAT gene expression and activities under low temperature stress [J]. The Journal of Applied Ecology (China), 2009, 20: 213−217.

[74] Gao Q H, Xu K, Wang X F, et al. Effect of grafting on cold tolerance in eggplant seedlings [J]. Acta Horticulturae,

2008, 771: 167-174.

[75] García-Sánchez F, Syvertsen J P, Gimeno V, et al. Responses to flooding and drought stress by two citrus rootstock seedlings with different water-use efficiency [J]. Physiologia Plantarum, 2007, 130: 532-542.

[76] Ghosh S, Mahoney S R, Penterman J N, et al. Ultrastructural and biochemical changes in chloroplasts during *Brassica napus* senescence [J]. Plant Physiology and Biochemistry, 2001, 39: 777-784.

[77] Gopal R, Rizvi A H. Excess lead alters growth, metabolism and translocation of certain nutrients in radish [J]. Chemosphere, 2008, 70: 1539-1544.

[78] Grover P, Rekhadevi P, Danadevi K, et al. Genotoxicity evaluation in workers occupationally exposed to lead [J]. International Journal of Hygiene and Environmental Health, 2010, 213: 99-106.

[79] Grzesiak S, Hura T, Grzesiak M T, et al. The impact of limited soil moisture and waterlogging stress conditions on morphological and anatomical root traits in maize (*Zea mays* L.) hybrids of different drought tolerance [J]. Acta Physiologiae Plantarum, 1999, 21: 305-315.

[80] Gupta D, Huang H, Yang X, et al. The detoxification of lead in *Sedum alfredii* H. is not related to phytochelatins but the glutathione [J]. Journal of Hazardous Materials, 2010, 177: 437-444.

[81] Gupta D, Nicoloso F, Schetinger M, et al. Antioxidant defense mechanism in hydroponically grown *Zea mays* seedlings under moderate lead stress [J]. Journal of

Hazardous Materials, 2009, 172: 479−484.

[82] Gutschick B P. Evolved strategies of nitrogen acquisition by plants [J]. The American Society of Naturalists, 1981, 118: 607−637.

[83] Guy C, Kaplan F, Kopka J, et al. Metabolomics of temperature stress [J]. Physiologia Plantarum, 2008, 132: 220−235.

[84] Hallik L, Niinemets Ü, Wright I J. Are species shade and drought tolerance reflected in leaf-level structural and functional differentiation in Northern Hemisphere temperate woody flora? [J]. New Phytologist, 2009, 184: 257−274.

[85] Hansen L D, Afzal M, Breidenbach R W, et al. High and low temperature limits to growth of tomato cells [J]. Planta, 1994, 195: 1−9.

[86] Harada T. Grafting and RNA transport via phloem tissue in horticultural plants [J]. Scientia Horticulturae, 2010, 125: 545−550.

[87] Hartley J, Cairney J W G, Freestone P, et al. The effects of multiple metal contamination on ectomycorrhizal Scots pine (*Pinus sylvestris*) seedlings [J]. Environmental Pollution, 1999, 106: 413−424.

[88] Hetherington A, Davis W J. Special issue: stomatal biology [J]. Journal of Experimental Botany, 1998, 49: 293−469.

[89] Hoek I H S, Hänischten C, Keijzer C J, et al. Development of the fifth leaf is indicative for whole plant performance at low temperature in tomato [J]. Annals of Botany, 1993, 72: 367−374.

[90] Hori Y, Arai K, Toki T. Studies on the effects of root temperature and its combination with air temperature on the growth and nutrition of vegetable crops. II. Carrot, celery, pepper, grafted cucumber and cucurbits used as stocks for cucumber (Japanese) [J]. Bulletin of the Horticultural Research Station, 1970, 9: 189−219.

[91] Hottes A C. Practical plant propagation: an exposition of the art and science of increasing plants as practiced by the nurseryman, florist and gardener [M]. New York: A. T. De La Mare Company, Inc, 1925.

[92] Huang B R, Nobel P S. Hydraulic conductivity and anatomy for lateral roots of Agave deserti during root-growth and drought-induced abscission [J]. Journal of Experimental Botany, 1992, 43: 1441−1449.

[93] Huang J Y, Lin C H. Cold water treatment promotes ethylene production and dwarfing in tomato seedlings [J]. Plant Physiology and Biochemistry, 2003, 41: 282−288.

[94] Hussain A, Black C R, Taylor I B, et al. Does an antagonistic relationship between ABA and ethylene mediate shoot growth when tomato (*Lycopersicon esculentum* Mill.) plants encounter compacted soil? [J]. Plant, Cell & Environment, 2000, 23: 1217−1226.

[95] Islam E, Liu D, Li T, et al. Effect of Pb toxicity on leaf growth, physiology and ultrastructure in the two ecotypes of *Elsholtzia argyi* [J]. Journal of Hazardous Materials, 2008, 154: 914−926.

[96] Islam E, Yang X, Li T, et al. Effect of Pb toxicity on root morphology, physiology and ultrastructure in the two

ecotypes of *Elsholtzia argyi* [J]. Journal of Hazardous Materials, 2007, 147: 806−816.

[97] Jefferies R. Cultivar responses to water stress in potato: effects of shoot and roots [J]. New Phytologist, 1993, 123: 491−498.

[98] Jiang W, Liu D. Pb-induced cellular defense system in the root meristematic cells of *Allium sativum* L [J]. BMC Plant Biology, 2010, 10: 40−40.

[99] Jones L H P, Clement C R, Hopper M J. Lead uptake from solution by perennial ryegrass and its transport from roots to shoots [J]. Plant Soil, 1973, 38: 403−414.

[100] Jongrungklang N, Toomsan B, Vorasoot N, et al. Rooting traits of peanut genotypes with different yield responses to pre-flowering drought stress [J]. Field Crops Research, 2011, 120: 262−270.

[101] Karlsson P S, Méndez M. The Resource Economy of Plant Reproduction [J]. Reproductive Allocation in Plants, 2005, 27: 1−49.

[102] Kasim W A. Physiological consequences of structural and ultra-structural changes induced by Zn stress in *Phaseolus vulgaris*. Ⅰ. Growth and photosynthetic apparatus [J]. International Journal of Botany, 2007, 1: 15−22.

[103] Kato C, Ohshima N, Kamada H, et al. Enhancement of the inhibitory activity for greening in xylem sap of squash root with waterlogging [J]. Plant Physiology and Biochemistry, 2001, 39: 513−519.

[104] Khah E M, Kakava E, Mavromatis A, et al. Effect of

grafting on growth and yield of tomato (*Lycopersicon esculentum* Mill.) in greenhouse and open-field [J]. Journal of Applied Horticulture, 2006, 8: 3—7.

[105] Kitao M, Lei T T, Koike T. Effects of manganese toxicity on photosynthesis of white birch (*Betula platyphyllavar. japonica*) seedlings [J]. Physiologia Plantarum, 1997, 101: 249—256.

[106] Kondo M, Murty M V R, Aragones D V. Characteristics of root growth and water uptake from soil in upland rice and maize under water stress [J]. SoilScience and Plant Nutrition, 2000, 46: 721—732.

[107] Kopittke P M, Asher C J, Kopittke R A, et al. Toxic effects of Pb^{2+} on growth of cowpea (*Vigna unguiculata*) [J]. Environmental Pollution, 2007, 150: 280—287.

[108] Kosobrukhov A, Knyazeva I, Mudrik V. Plantago major plants responses to increase content of lead in soil: growth and photosynthesis [J]. Plant Growth Regulation, 2004, 42: 145—151.

[109] Landberg T, Greger M. Differences in uptake and tolerance to heavy metals in *Salix* from unpolluted and polluted areas [J]. Applied Geochemistry, 1996, 11: 175—180.

[110] Lane S D, Martin E S. A histochemical investigation of lead uptake in *Raphanus sativus* [J]. New Phytologist, 1977, 79: 281—286.

[111] Laporte M M, Delph L F. Sex-specific physiology and source-sink relations in the dioecious plant *Silene latifolia* [J]. Oecologia, 1996, 106: 63—72.

[112] Lauteri M, Scartazza A, Ouido M C, et al. Genetic variation in photosynthetic capacity, carbon isotope discrimination and mesophyll conductance in provenances of *Castanea sativa* adapted to different environments [J]. Functional Ecology, 1997, 11: 675−683.

[113] Lee J M, Oda M. Grafting of herbaceous vegetable and ornamental crops [J]. Horticultural Reviews, 2003, 28: 61−124.

[114] Lee S H, Ahn S J, Im Y J, et al. Differential impact of low temperature on fatty acid unsaturation and lipoxygenase activity in figleaf hourd and cucumber roots [J]. Biochemical and Biophysical Research Communications, 2005, 330: 1194−1198.

[115] Lee S H, Chung G C, Steudle E. Gating of aquaporins by low temperature in roots of chilling-sensitive cucumber and chilling-tolerant figleaf gourd [J]. Journal of Experimental Botany, 2005, 56 (413): 985−995.

[116] Lee S H, Chung G C, Steudle E. Low temperature and mechanical stresses differently gate aquaporins of root cortical cells of chilling-sensitive cucumber and-resistant figleaf gourd [J]. Plant, Cell & Environment, 2005, 28: 1191−1202.

[117] Lee S H, Singh A P, Chung G C, et al. Exposure of roots of cucumber (*Cucumis sativus*) to low temperature severely reduces root pressure, hydraulic conductivity and active transport of nutrients [J]. Physiologia Plantarum, 2004, 120 (3): 413−420.

[118] Lee S H, Singh A P, Chung G C. Rapid accumulation

of hydrogen peroxide in cucumber roots due to exposure to low temperature appears to mediate decreases in water transport [J]. Journal of Experimental Botany, 2004, 55: 1733-1741.

[119] Lee S H, Zwiazek J J, Chung G C. Light-induced transpiration alters cell water relations in figleaf gourd (*Cucurbita ficifolia*) seedlings exposed to low root temperatures [J]. Physiologia Plantarum, 2008, 133: 354-362.

[120] Leigh A, Cosgrove M J, Nicotra A B. Reproductive allocation in a gender dimorphic shrub: anomalous female investment in *Gynatrix pulchella*? [J]. Journal of Ecology, 2006, 94: 1261-1271.

[121] LeNoble M E, Spollen W G, Sharp R E. Maintenance of shoot growth by endogenous ABA: genetic assessment of the involvement of ethylene suppression [J]. Journal of Experimental Botany, 2004, 55: 237-245.

[122] Levins R. Evolution in changing environments [M]. Princeton: Princeton University Press, 1968.

[123] Li C Y, Berninger F, Koskela J, et al. Drought responses of *Eucalyptus microtheca* provenances depend on seasonality of rainfall in their place of origin [J]. Australian Journal of Plant Physiology, 2000, 27: 231-238.

[124] Li C Y, Yin C Y, Liu S R. Different responses of two contrasting *Populus davidiana* populations to exogenous abscisic acid application [J]. Environmental and Experimental Botany, 2004, 51: 237-246.

[125] Li C. Variation of seedling traits of *Eucalyptus microtheca* origins in different water regimes [J]. Silvae Genetica, 1998, 47: 132−136.

[126] Li T, Yu X. Effect of Cu^{2+}, Zn^{2+}, and Mn^{2+} on SOD activity of cucumber leaves extraction after low temperature stress [J]. Acta Horticulturae Sinica, 2007, 34: 895−900.

[127] Li Y T, Tian H X, Li X G, et al. Higher chilling-tolerance of grafted-cucumber seedling leaves upon exposure to chilling stress [J]. Agricultural Science China, 2008, 7: 570−576.

[128] Lian H L, Yu X, Ye Q, et al. The role of aquaporin RWC3 in drought avoidance in rice [J]. Plant and Cell Physiology, 2004, 45: 481−489.

[129] Liao C T, Lin C H. Photosynthetic response of grafted bitter melon seedling to flood stress [J]. Environmental and Experimental Botany, 1996, 36: 167−172.

[130] Lichtenthaler H K. Chlorophylls and carotenoids: pigments of photosynthetic biomembranes [J]. Methods in Enzymology, 1987, 148: 350−382.

[131] Liu D, Li T, Jin X, et al. Lead induced changes in the growth and antioxidant metabolism of the lead accumulating and non-accumulating ecotypes of *Sedum alfredii* [J]. Journal of Integrative Plant Biology, 2008, 50: 129−140.

[132] Lloyd D G. Selection of combined versus separate sexes in seed plants [J]. The American Naturalist, 1982, 120: 571−585.

[133] Lux A, Šottníková A, Opatrná J, et al. Differences in structure of adventitious roots in *Salix* clones with contrasting characteristics of cadmium accumulation and sensitivity [J]. Physiologia Plantarum, 2004, 120: 537—545.

[134] Maestri E, Marmiroli M, Visioli G, et al. Metal tolerance and hyperaccumulation: costs and trade-offs between traits and environment [J]. Environmental and Experimental Botany, 2010, 68: 1—13.

[135] Masuda M, Gomi K. Diurnal change of the exudation rate and the mineral concentration in xylem sap after decapitation of grafted and non-grafted cucumbers [J]. Journal of the Japanese Society for Horticultural Science, 1982, 51: 293—298.

[136] Masuda M, Gomi K. Mineral absorption and oxygen consumption in grafted and non-grafted cucumbers [J]. Journal of the Japanese Society for Horticultural Science, 1984, 54: 414—419.

[137] Matsui T, Singh B B. Root characteristics in cowpea related to drought tolerance at the seedling stage [J]. Experimental Agriculture, 2003, 39: 29—38.

[138] Matsuyama S, Sakimoto M. Allocation to reproduction and relative reproductive costs in two species of dioecious Anacardiaceae with contrasting phenology [J]. Annals of Botany, 2008, 101: 1391—1400.

[139] Maustakas M, Ouzounidou G, Symeonidis L, et al. Field study of the effects of copper on wheat photosynthesis and productivity [J]. Soil Science &

Plant Nutrition, 1997, 43: 531−539.

[140] Maxwell K, Johnson G N. Chlorophyll fluorescence-a practical guide [J]. Journal of Experimental Botany, 2000, 51: 659−668.

[141] McLean E H, Ludwig M, Grierson P F. Root hydraulic conductance and aquaporin abundance respond rapidly to partial root-zone drying events in a riparian *Melaleuca* species [J]. New Phytologist, 2011, 192: 664−675.

[142] McMichael B L, Quisenberry J E. Genetic variation for root-shoot relationships among cotton germplasm [J]. Environmental and Experimental Botany, 1991, 31: 461−470.

[143] Menhenett R, Wareing P F. Possible involvement of growth substances in the response of tomato plants (*Lycopersicon esculentum* Mill.) to different soil temperatures [J]. The Journal of Horticultural Science & Biotechnology, 1975, 50: 381−397.

[144] Miao M, Zhang Z, Xu X, et al. Different mechanisms to obtain higher fruit growth rate in two cold-tolerant cucumber (*Cucumis sativus* L.) lines under low night temperature [J]. Scientia Horticulturae, 2009, 119: 357−361.

[145] Mishra S, Srivastava S, Tripathi R, et al. Lead detoxification by coontail (*Ceratophyllum demersum* L.) involves induction of phytochelatins and antioxidant system in response to its accumulation [J]. Chemosphere, 2006, 65: 1027−1039.

[146] Miyazawa S I, Yoshimura S, Shinzaki Y, et al.

Deactivation of aquaporins decreases internal conductance to CO_2 diffusion in tobacco leaves grown under long-term drought [J]. Functional Plant Biology, 2008, 35: 553—564.

[147] Mohan B S, Hosetti B B. Potential phytotoxicity of lead and cadmium to *Lemnaminor* grown in sewage stabilization ponds [J]. Environmental Pollution, 1997, 98: 233—238.

[148] Monclus R, Dreyer E, Villar M, et al. Impact of drought on productivity and water use efficiency in 29 genotypes of *Populus deltoides* × *Populus nigra* [J]. New Phytologist, 2006, 169: 765—777.

[149] Morison J I L, Morecroft M D. Plant growth and climate change [M]. Oxford: Blackwell Publishing, 2006.

[150] Munne-Bosch S, Mueller M, Schwarz K, et al. Diterpenes and antioxidative protection in drought-stressed *Salvia officinalis* plants [J]. Journal of Plant Physiology, 2001, 158: 1431—1437.

[151] Nagel K A, Kastenholz B, Jahnke S, et al. Temperature responses of roots: impact on growth, root system architecture and implications for phenotyping [J]. Functional Plant Biology, 2009, 36: 947—959.

[152] Newman M C, Unger M A. Fundamentals of Ecotoxicology Lewis Publishers [J]. Boca Raton, Florida, 2003, 458.

[153] Nieuwhof M, Keizer L C P, Zijlstra S, et al. Genotypic variation for root activity in tomato (*Lycopersicon esculentum* Mill.) at different root temperatures [J]. Journal of Genetics and Breeding, 1999, 53: 271—278.

[154] North G B, Nobel P S. Changes in hydraulic conductivity and anatomy caused by drying and rewetting roots of *Agave deserti* (Agavaceace) [J]. American Journal of Botany, 1991, 78: 906−915.

[155] North G B, Nobel P S. Water uptake and structural plasticity along roots of a desert succulent during prolonged drought [J]. Plant, Cell & Environment, 1998, 21: 705−713.

[156] Okimura M, Matso S, Arai K, et al. Influence of soil temperature on the growth of fruit vegetable grafted on different stocks [J]. Bulletin of the Vegetable and Ornamental Crops Research Station. Series C. Kurume (Japan), 1986, 9: 43−58.

[157] Olsson T, Leverenz J W. Non-uniform stomatal closure and the apparent convexity of the photosynthetic photon flux density response curve [J]. Plant, Cell & Environment, 1994, 17: 701−710.

[158] Oncel I, Keleş Y, Ustün A S. Interactive effects of temperature and heavy metal stress on the growth and some biochemical compounds in wheat seedlings [J]. Environmental Pollution, 2000, 107: 315−20.

[159] Parys E, Romanowska E, Siedlecka M, et al. The effect of lead on photosynthesis and respiration in detached leaves and in mesophyll protoplasts of *Pisum sativum* [J]. Acta Physiologiae Plantarum, 1998, 20: 313−322.

[160] Passioura J B. Drought and drought tolerance [J]. Plant Growth Regulation, 1996, 20: 79−83.

[161] Passioura J B. Environmental biology and crop improvement [J]. Functional Plant Biology, 2002, 29: 537-546.

[162] Paul M J, Foyer C H. Sink regulation of photosynthesis [J]. Journal of Experimental Botany, 2001, 52: 1383-1400.

[163] Penuelas J, Filella I. Metal pollution in Spanish terrestrial ecosystems during the twentieth century [J]. Chemosphere, 2002, 46: 501-505.

[164] Perez-Alfocea F, Albacete A, Ghanem M E, et al. Hormonal regulation of source-sink relations to maintain crop productivity under salinity: a case study of root-to-shoot signalling in tomato [J]. Functional Plant Biology, 2010, 37: 592-603.

[165] Pogany M, Elstner E F, Barna B. Cytokinin gene introduction confers tobacco necrosis virus resistance and higher antioxidant levels in tobacco [J]. Free Radical Research, 2003, 37: 15-16.

[166] Porra R J, Thompson W A, Kriedmann P E. Determination of accurate extinction coefficients and simultaneous equations for assaying chlorophylls a and b extracted with four different solvents: verification of the concentration of chlorophyll standards by atomic absorption spectroscopy [J]. BBA-Bioenergetics, 1989, 975: 384-394.

[167] Poschenrieder C H, Barceló J. Water relations in heavy metal stressed plants [M] //Prasad M N V, Hagemeyer J. Heavy metal stress in plants. Heidelberg: Springer-

Verlag, 1999: 207-229.

[168] Prasad M N V, Strzalka K. Impact of heavy metals on photosynthesis [M] //Prasad M N V, Hagemeyer J. Heavy metal stress in plants. Heidelberg: Springer-Verlag, 1999: 117-137.

[169] Prioul J L, Chartier P. Partitioning of transfer and carboxylation components of intracellular resistance to photosynthetic CO_2 fixation: a critical analysis of the methods used [J]. Annals of Botany-London, 1977, 41: 789-800.

[170] Qin L, He J, Lee S K, et al. An assessment of the role of ethylene in mediating lettuce (*Lactuca sativa*) root growth at high temperatures [J]. Journal of Experimental Botany, 2007, 58: 3017-3024.

[171] Qufei L, Fashui H. Effects of Pb^{2+} on the Structure and Function of Photosystem II of *Spirodela polyrrhiza* [J]. Biological Trace Element Research, 2009, 129: 251-260.

[172] Qureshi M, Abdin M, Qadir S, et al. Lead-induced oxidative stress and metabolic alterations in *Cassia angustifolia* [J]. Biologia Plantarum, 2007, 51: 121-128.

[173] Ranathunge K, Lin J X, Steudle E, et al. Stagnat deoxygenated growth enhances root suberization and lignifications, but differentially affects water and NaCl permeabilities in rice (*Oryza sativa* L.) roots [J]. Plant, Cell & Environment, 2011, 34: 1223-1240.

[174] Rascio N, Dalla Vecchia F, Ferretti M, et al. Some

effects of cadmium on maize plants [J]. Archives of Environmental Contamination and Toxicology, 1993, 25: 244−249.

[175] Reich P, Hinckley T. Influence of pre-dawn water potential and soil-to-leaf hydraulic conductance on maximum daily leaf diffusive conductance in two oak species [J]. Functional Ecology, 1989, 3: 719−726.

[176] Renner S S, Ricklefs R E. Dioecy and its correlates in the flowering plants [J]. American Journal of Botany, 1995, 82: 596−606.

[177] Rhee J Y, Lee S H, Singh A P, et al. Detoxification of hydrogen peroxide maintains the water transport activity in figleaf gourd (*Cucurbita ficifolia*) root system exposed to low temperature [J]. Physiologia Plantarum, 2007, 130: 177−184.

[178] Richards A J. Plant breeding systems [M]. Massachusetts: George Allen and Unwin Ltd, 1986.

[179] Rivero R M, Kojima M, Gepstein A, et al. Delayed leaf senescence induces extreme drought tolerance in a flowering plant [J]. Proceedings of the National Academy of Sciences, 2007, 104: 19631−19636.

[180] Rivero R M, Ruiz J M, Romero L. Can grafting in tomato plants strengthen resistance to thermal stress? [J]. Journal of the Science of Food and Agriculture, 2003, 83: 1315−1319.

[181] Rivero R M, Ruiz J M, Sanchez E, et al. Does grafting provide tomato plants an advantage against H_2O_2 production under conditions of thermal shock? [J].

Physiologia Plantarum, 2003, 117: 44−50.

[182] Robinson B H, Mills T M, Petit D, et al. Natural and induced cadmium-accumulation in poplar and willow: implications for phytoremediation [J]. Plant Soil, 2000, 227: 301−306.

[183] Romanowska E, Igamberdiev A U, Parys E, et al. Stimulation of respiration by Pb^{2+} in detached leaves and mitochondria of C_3 and C_4 plants [J]. Physiologia Plantarum, 2002, 116: 148−154.

[184] Romanowska E, Pokorska B, Siedlecka M. The effects of oligomycin on content of adenylates in mesophyll protoplasts, chloroplasts and mitochondria from Pb^{2+} treated pea and barley leaves [J]. Acta Physiologiae Plantarum, 2005, 27: 29−36.

[185] Romanowska E, Wróblewska B, Drozak A, et al. High light intensity protects photosynthetic apparatus of pea plants against exposure to lead [J]. Plant Physiology and Biochemistry, 2006, 44: 387−394.

[186] Rouphael Y, Cardarelli M, Colla G, et al. Yield, mineral composition, water relations, and water use efficiency of grafted mini-watermelon plants under deficit irrigation [J]. Hortscience, 2008, 43: 730−736.

[187] Ruiz-Lozano J M, Alguacil M M, Bárzana G, et al. Exogenous ABA accentuates the differences in root hydraulic properties between mycorrhizal and nonmycorrhizal maize plants through regulation of PIP aquaporins [J]. Plant Molecular Biology, 2009, 70: 565−579.

[188] Sanders P L, Markhart A H. Interspecific grafts

demonstrate root system control of leaf water status in water stressed Phaseolus [J]. Journal of Experimental Botany, 1992, 43: 1563—1567.

[189] Satisha J, Prakash G S, Bhatt R M, et al. Physiological mechanisms of water use efficiency in grape rootstocks under drought conditions [J]. International Journal of Agricultural Research, 2007, 2: 159—164.

[190] Schwab K B, Schreiber U, Heber U. Response of photosynthesis and respiration of resurrection plants to desiccation and rehydration [J]. Planta, 1989, 177: 217—227.

[191] Schwartz M D. Advancing to full bloom: planning phenological research for the twenty first century [J]. International Journal of Biometeorology, 1999, 42: 113—118.

[192] Serraj R, Sinclair T R. Processes contributing to N_2-fixation insensitivity to drought in the soybean cultivar Jackson [J]. Crop Science, 1996, 36: 961—968.

[193] Shahid M, Pinelli E, Pourrut B, et al. Lead-induced genotoxicity to *Vicia faba* L. roots in relation with metal cell uptake and initial speciation [J]. Ecotoxicology and Environmental Safety, 2011, 74: 78—84.

[194] Sharp R E, LeNoble M E. ABA, ethylene and the control of shoot and root growth under water stress [J]. Journal of Experimental Botany, 2002, 53: 33—37.

[195] Shine R. Ecological causes for the evolution of sexual dimorphism: A review of the evidence [J]. The Quarterly Review of Biology, 1989, 64: 419—461.

[196] Singh R, Tripathi R D, Dwivedi S, et al. Lead bioaccumulation potential of an aquatic macrophyte Najas indica is related to antioxidant system [J]. Bioresource Technology, 2010, 101: 3025-3032.

[197] Spollen W G, LeNoble M E, Samuels T D, et al. Abscisic acid accumulation maintains maize primary root elongation at low water potentials by restricting ethylene production [J]. Plant Physiology, 2000, 122: 967-976.

[198] Steinmuller D, Tevini M. Composition and function of plastoglobuli, I. Isolation and purification from chloroplasts and chromoplasts [J]. Planta, 1985, 163: 201-207.

[199] Steudle E. Water uptake by root: effects of water deficit [J]. Journal of Experimental Botany, 2000, 51: 1531-1542.

[200] Subrahmanyam D, Rathore V S. Influence of manganese toxicity on photosynthesis in ricebean (*Vigna umbellata*) seedlings [J]. Photosynthetica, 2000, 38: 449-453.

[201] Tachibana S. Comparison of effects of root temperature on the growth and mineral nutrition of cucumber and figleaf gourd [J]. Journal of the Japanese Society for Horticultural Science, 1982, 51: 299-308.

[202] Tachibana S. Effect of root temperature on the rate of water and nutrient absorption in cucumber and figleaf gourd [J]. Journal of the Japanese Society for Horticultural Science, 1987, 55: 461-467.

[203] Tan L P, He J, Lee S K. Effects of root-zone

temperature on the root development and nutrient uptake of *Lactuca sativa* L. cv. Panama grown in an aeroponic system in the tropics [J]. Journal of Plant Nutrition, 2002, 25: 297—314.

[204] Tang A C, Kawamitsu Y, Kanechi M, et al. Photosynthetic oxygen evolution at low water potential in leaf discs lacking an epidermis [J]. Annals of Botany, 2002, 89: 861—870.

[205] Tezara W, Mitchell V J, Driscoll S D, et al. Water stress inhibits plant photosynthesis by decreasing coupling factor and ATP [J]. Nature, 1999, 401: 914—917.

[206] Tezara W, Mitchell V J, Driscoll S P, et al. Effects of water deficit and its interaction with CO_2 supply on the biochemistry and physiology of photosynthesis in sunflower [J]. Journal of Experimental Botany, 2001, 53: 1781—1791.

[207] Uzu G, Sobanska S, Aliouane Y, et al. Study of lead phytoavailability for atmospheric industrial micronic and sub-micronic particles in relation with lead speciation [J]. Environmental Pollution, 2009, 157: 1178—1185.

[208] Uzu G, Sobanska S, Sarret G, et al. Foliar lead uptake by lettuce exposed to atmospheric fallouts [J]. Environmental Science & Technology, 2010, 44: 1036—1042.

[209] Van der Ploeg A, Heuvelink E, Venema J H. Wild relatives as a source for sub-optimal temperature tolerance in tomato [J]. Acta Horticulturae, 2007, 761: 127—133.

[210] van Kooten O, Snel J F H. The use of chlorophyll fluorescence nomenclature in plant stress physiology [J]. Photosynthesis Research, 1990, 25: 147−150.

[211] Venema J H, Dijk B E, Bax J M, et al. Grafting tomato (*Solanum lycopersicum*) onto the rootstock of a high-altitude accession of *Solanum habrochaites* improves suboptimal-temperature tolerance [J]. Environmental and Experimental Botany, 2008, 63 (1−3): 359−367.

[212] Venema J H, Linger P, van Heusden A W, et al. The inheritance of chilling tolerance in tomato (*Lycopersicon* spp.) [J]. Plant Biology, 2005, 7: 118−130.

[213] Venema J H, Posthumus F, Van Hasselt P R. Impact of suboptimal temperature on growth, photosynthesis, leaf pigments, and carbohydrates of domestic and high aötitude, wild *Lycopersicon* species [J]. Journal of Plant Physiology, 1999, 155: 711−718.

[214] Veselova S V, Farhutinov R G, Veselov S Y, et al. The effect of root cooling on hormone content, leaf conductance and root hydraulic conductivity of durum wheat seedlings (*Triticum durum* L.) [J]. Journal of Plant Physiology, 2005, 16: 21−26.

[215] Wang W, Vinocur B, Altman A. Plant responses to drought, salinity and extreme temperatures: towards genetic engineering for stress tolerance [J]. Planta, 2003, 218: 1−14.

[216] Ward J T, Lahner B, Yakubova E, et al. The effect of iron on the primary root elongation of arabidopsis during phosphate deficiency [J]. Plant Physiology, 2008,

147: 1181−1191.

[217] Warren C R, Adams M A. Internal conductance does not scale with photosynthetic capacity: implications for carbon isotope discrimination and the economics of water and nitrogen use in photosynthesis [J]. Plant, Cell & Environment, 2006, 29: 192−201.

[218] Wikbergi J, Ogreni E. Variation in drought resistance, drought acclimation and water conservation in four willow cultivars used for biomass production [J]. Tree Physiology, 2007, 27: 1339−1346.

[219] Wilkins D A. A technique for the measurement of lead tolerance in plants [J]. Nature, 1957, 180: 37−38.

[220] Wilkins D A. The measurement of tolerance to edaphic factors by means of root growth [J]. New Phytologist, 1978, 80: 623−633.

[221] Wu Y J, Cosgrove D J. Adaptation of roots to low water potentials by changes in cell wall extensibility and cell wall proteins [J]. Journal of Experimental Botany, 2000, 51: 1543−1553.

[222] Xu X, Peng G Q, Wu C C, et al. Drought inhibits photosynthetic capacity more in females than in males of *Populus cathayana* [J]. Tree Physiology, 2008, 28: 1751−1759.

[223] Xu X, Yang F, Xiao X W, et al. Sex-specific responses of Populus cathayana to drought and elevated temperatures [J]. Plant, Cell & Environment, 2008, 31: 850−860.

[224] Yan Z Z, Ke L, Tam N F Y. Lead stress in seedlings of

Avicennia marina, a common mangrove species in South China, with and without cotyledons [J]. Aquatic Botany, 2010, 92: 112−118.

[225] Yetisir H, Caliskan M E, Soylu S, et al. Some physiological and growth responses of watermelon [*Citrullus lanatus* (Thunb.) Matsum. and Nakai] grafted onto *Lagenaria siceraria* to flooding [J]. Environmental and Experimental Botany, 2006, 58: 1−8.

[226] Zayed A, Gowthaman S, Terry N. Phytoaccumulation of trace elements by wetland plants: I. Duckweed [J]. Journal of Environmental Quality, 1998, 27: 715−721.

[227] Zhang S, Chen F G, Peng S M, et al. Comparative physiological, ultrastructural and proteomic analyses reveal sexual differences in the responses of *Populus cathayana* under drought stress [J]. Proteomics, 2010, 10: 2661−2677.

[228] Zhang W, Cai Y, Tu C, et al. Arsenic speciation and distribution in an arsenic hyperaccumulating plant [J]. Science of the Total Environment, 2002, 300: 167−177.

[229] Zhao H X, Li Y, Duan B L, et al. Sex-related adaptive responses of *Populus cathayana* to photoperiod transitions [J]. Plant, Cell & Environment, 2009, 32: 1401−1411.

[230] Zhou Y H, Huang L F, Zhang Y, et al. Chill-induced decrease in capacity of RuBP carboxylation and associated

H_2O_2 accumulation in cucumber leaves are alleviated by grafting onto Figleaf Gourd [J]. Annals of Botany, 2007, 100: 839—848.

[231] Zhou Y H, Lam H M, Zhang J H. Inhibition of photosynthesis and energy dissipation induced by water and high light stresses in rice [J]. Journal of Experimental Botany, 2007, 58: 1207—1217.

[232] Zhou Y H, Wu J X, Zhu L J, et al. Effects of phosphorus and chilling under low irradiance on photosynthesis and growth of tomato plants [J]. Biologia Plantarum, 2009, 53 (2): 378—382.

[233] Zhou Y H, Yu J Q, Huang L F, et al. The relationship between CO_2 assimilation, photosynthetic electron transport and water-water cycle in chillexposed cucumber leaves under low light and subsequent recovery [J]. Plant, Cell & Environment, 2004, 27: 1503—1514.

[234] Zhou Y H, Yu J Q, Mao W H, et al. Genotypic variation of rubisco expression, photosynthetic electron flow and antioxidant metabolism in the chloroplasts of chill-exposed cucumber plants [J]. Plant and Cell Physiology, 2006, 47: 192—199.

[235] Zijlstra S, den Nijs A P M. Effects of root systems of tomato genotypes on growth and earliness, studied in grafting experiments at low temperature [J]. Euphytica, 1987, 36: 693—700.

[236] Zimmerman J K, Lechowicz M J. Responses to moisture stress in male and female plants of *Rumex acetosella* 1. (Polygonaceae) [J]. Oecologia, 1982, 53: 305—309.

[237] Zollinger N, Kjelgren R, Cerny-Koenig T, et al. Drought responses of six ornamental herbaceous perennials [J]. Scientia Horticulturae, 2006, 109: 267−274.

[238] Zu Y Q, Li Y, Chen J J, et al. Hyperaccumulation of Pb, Zn, and Cd in herbaceous grown on lead-zinc mining area in Yunnan, China [J]. Environment International, 2005, 31: 755−762.